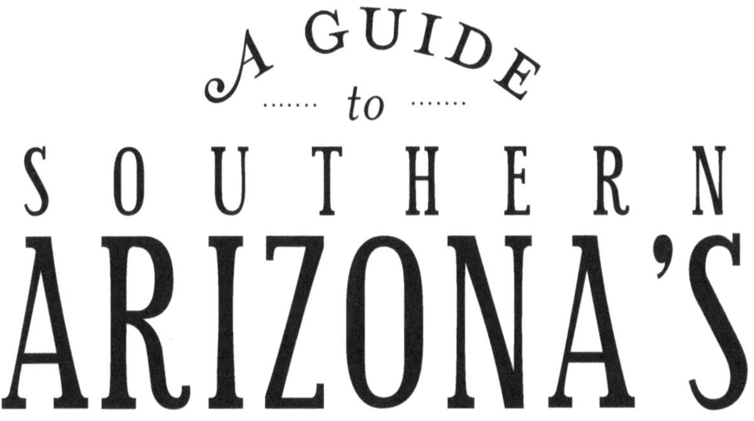

Historic Farms & Ranches

A Guide to Southern Arizona's Historic Farms & Ranches

Rustic Southwest Retreats

Lili DeBarbieri

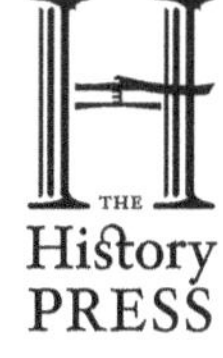

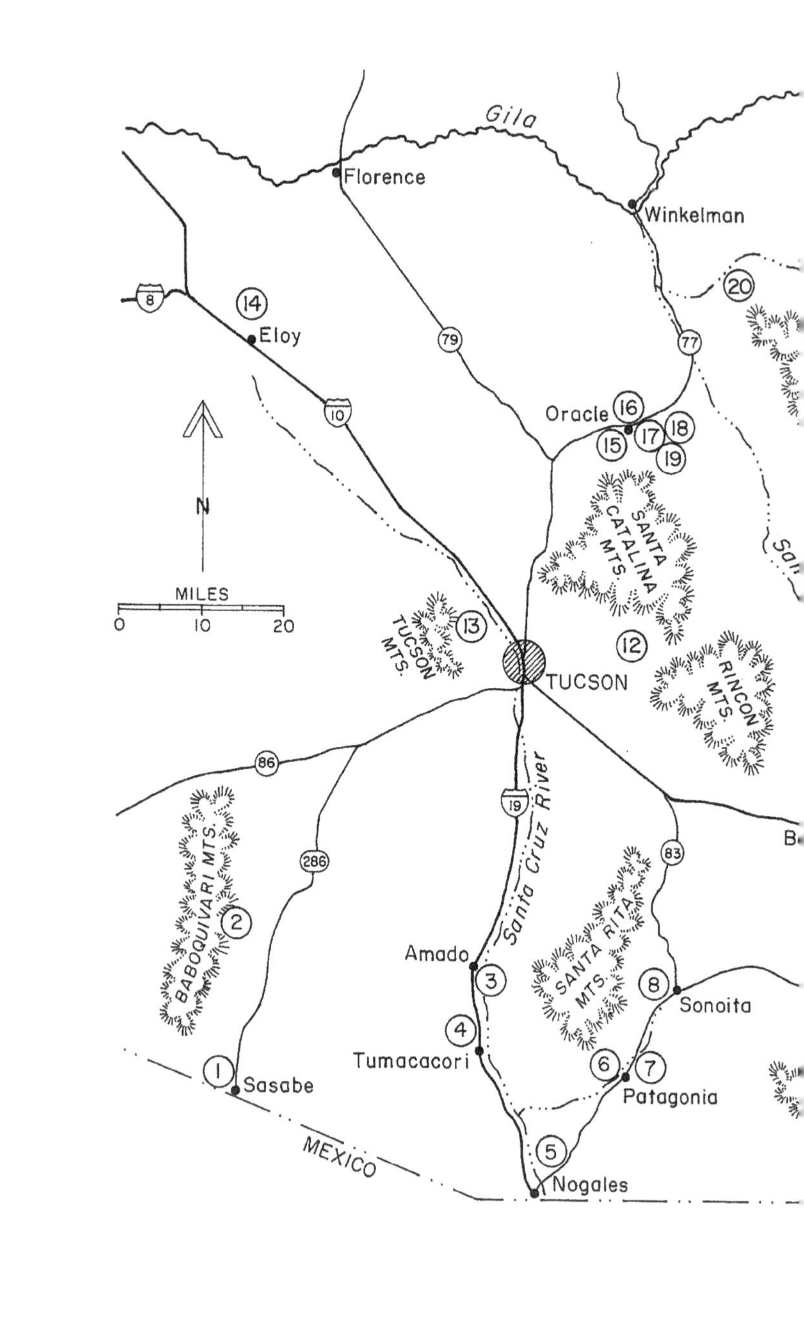

Gila
Florence
Winkelman
14
Eloy
8
79
10
77
20
Oracle
16
15
17
18
19
N
MILES
0
10
20
SANTA CATALINA MTS.
TUCSON MTS.
13
12
TUCSON
RINCON MTS.
86
19
Santa Cruz River
83
286
BABOQUIVARI MTS.
2
SANTA RITA MTS.
Amado
3
8
Sonoita
4
Tumacacori
6
7
Patagonia
1
Sasabe
5
MEXICO
Nogales

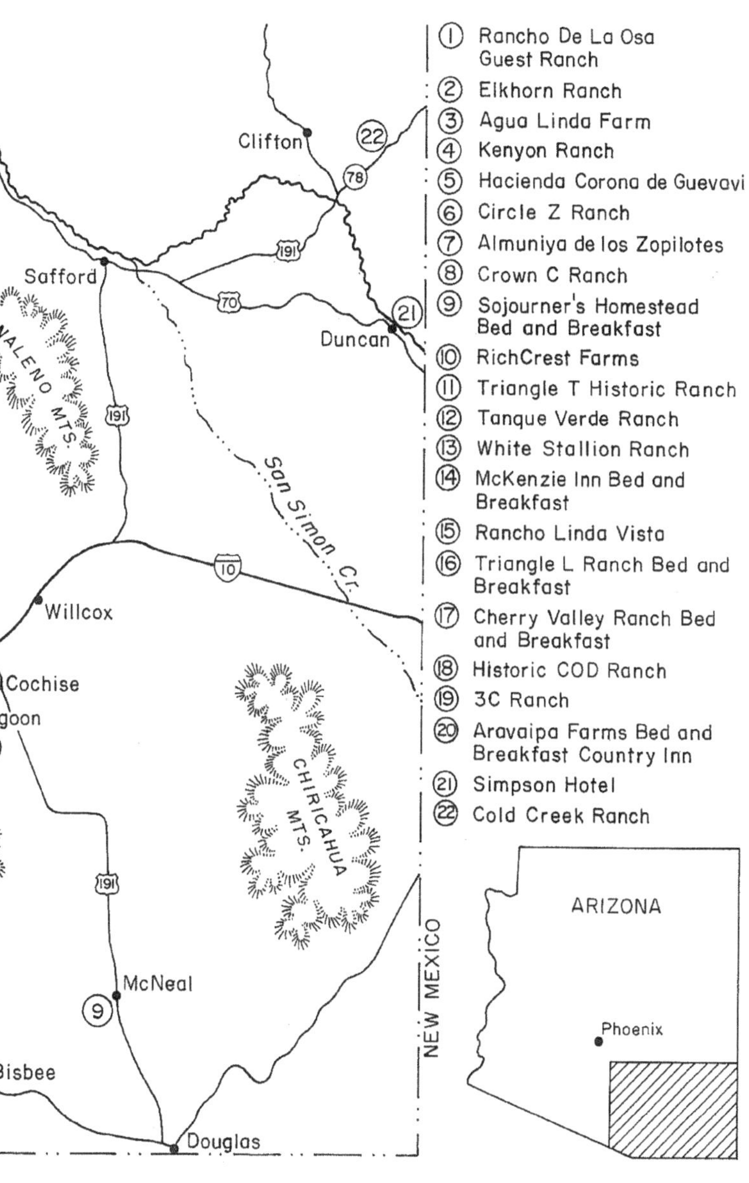

Map of Southern Arizona. *Drawn by Chuck Sternberg.*

Published by The History Press
Charleston, SC 29403
www.historypress.net

Cover image: Norma Basset Hall, *Untitled (Arizona Landscape)*, c. 1949, watercolor (detail).
Collection of the Tucson Museum of Art.
Photo Credit: Wilson Graham Photography

First published 2012

Manufactured in the United States

ISBN 978.1.60949.460.5

Library of Congress CIP data applied for.

Contents

Acknowledgements

I would like to thank the following individuals and organizations for their invaluable assistance, expertise and support, without which this book could not have been written: Farm Stay U.S., Local First Arizona, the Arizona Farm Bureau, the Dude Ranchers' Association, the Arizona Cooperative Extension, the Arizona Historical Society, the Oracle Historical Society, the Tubac Historical Society, World Wide Opportunities on Organic Farms (WWOOF), Native Seeds/SEARCH, the Tucson Museum of Art, Aubrie Koenig, Marshall Trimble, Scottie Jones, Grace Berg, Susan Dolan, Russell Tronstad, Michael and Pani McKenzie, Veronica Schultz, Bob Cote, Wendy Stover, Mary Miller, Carol Steele, Russell True, Susan Woodruff, Patricia Woodruff, Jane Woodruff, Stacy Raneri, Chuck Sternberg, Judy Stewart, Sharon Holnback, Eric Schwennesen, Deborah Mendelsohn, Sherry Harrison, Linda Kelley, Stewart and Laurel Loew, Vera Weintraub, Lynne Thompson Rugg, Joyce Winough, Jock Soper, Diana Nash, Fran Karanovich, Kimber Lanning, Jim Dumas, Adrian Darimont, Steve Malkin, Sidney Franklin, Colleen Hodson, Sarah Potenza, Julie Murphree, Amanda Rockafellow Photography, Rosemary Woods, Jeff Smith, Ann Haver Allen, Gary Paul Nabhan, Tim Tracy, Terri Allen, Charles and Barbara Findeisen and Theodore G. Manno.

Introduction

Southern Arizona's light and landscape are awe inspiring. The desert is a majestic palate of blues, pinks, purples, golds and subtle greens. At night, it is also alive with sound. Like a Hollywood movie set, diverse ecologies of high and low desert, forests, mountains and grasslands are the setting where the stories of the following historic ranches and farms are enacted.

Although Arizona is more well-known for its golf and resort-related recreation, agriculture quietly plays a major role in the state's business and cultural arenas. This is famously the land of the 5 Cs—cattle, copper, cotton, citrus and climate. In this year of Arizona's Centennial—one hundred years of statehood—it is important to note that ranches and farms were the state's earliest form of commerce, its residents' earliest source of livelihood and one of the earliest forms of tourism.

Although no two are exactly alike, these ranches and farms were shaped by the same historical circumstances, and through their unique stories, the common bonds, originality and diversity that make up this area of the American Southwest can be glimpsed.

Agricultural tourism can take many forms—farmers' markets, pick-your-own farms, restaurants and tours. Instead, I chose to focus on farms and ranches that offer overnight accommodations in addition to significant community involvement and environmental friendliness.

Hacienda, bed-and-breakfast, inn, ranch stay, farm stay, guest ranch, dude ranch and ranch resort—all these terms have been used to describe the properties in this book. With an intangible spirit that defies boxed-in definitions, it is the personal and inspirational stories and interactions guests have with this type of tourism that are its real distinguishing feature. First and foremost, the farms and ranches are people's homes. When staying on a farm or ranch, you meet and interact with sophisticated owners, management and staff that have an incredible amount of passion, interest, know-how and personal connection to their work. They will tell you the history of their land and family. You will eat with the property's owners, learn from them, be entertained by them and possibly even work beside them. In contrast to other vacation properties, farms and ranches inspire passionate emotions from their owners, returnee guests and the public.

A stay on a guest ranch or farm may not be for the mainstream vacationer. Yet they have a very wide appeal—history and film buffs; nature and outdoor enthusiasts; and art, literature and culture aficionados all have ample opportunities for enjoyment at these locations. These exceptional places offer recreation, modern day educational experiences, glimpses into Southern Arizona's fascinating past and a respite from the suffocating concrete that is a large part of the daily grind.

As an avid agri-tourist and former guest ranch employee, I wondered what this state had to offer in that realm. The idea for the book that was born is best described as an amalgam of all my personal interests—travel, history, food, art, conservation, wildlife and especially the Southwest. I hope you have the pleasure of finding out firsthand what makes the following places and people so unforgettable.

Chapter 1

The Bountiful Agricultural Tourism of Southern Arizona

"Simply put, agriculture is growing things on purpose."
—*R. Douglas Hurt*

From scenic vineyards and ranchlands to pistachio orchards, chili pepper fields and pumpkin patches, the vast diversity of agriculture in Southern Arizona makes for excellent opportunities for agricultural tourism. Agricultural tourism, or "agri-tourism," is businesses conducted by working farms or ranches for the attraction, enjoyment and education of visitors. Agri-tourism activities include visits to wineries, pick-your-own farms, roadside stands, farm or ranch bed-and-breakfasts, guest ranches, agricultural festivals, fairs and farm tours.

Through agri-tourism, travelers learn about agriculture and, in turn, support farms economically through their direct or indirect purchases. In contrast to other tourists, agricultural tourists have a greater propensity for returning to one location year after year. Although by definition, agri-tourists are generally more interested in healthy living and the environment than other kinds of travelers, there are subtle differences between the types of tourists that stay on ranches and farms. Guest ranch vacationers and retreaters tend to stay longer, travel in larger groups, have slightly higher income and are older in age, in some cases. They are drawn by nostalgia and the romance of the West. Farm stay guests, in contrast, are singles or couples that may overnight on a farm for a long weekend, are more food-oriented and drawn to the "country life" in a wider sense.

Even with Southern Arizona's large concentration of farms and ranches, many still have "no idea how great the area's agri-tourism really is," says Kimber Lanning, executive director of Local First Arizona. "Most people cannot believe they are in Arizona. The landscape looks like the countryside in Italy or France."

Arizona's agriculture and tourism go hand in hand as "both depend in large part on nature, the weather and the quality of Arizona's natural resources," writes Russell Tronstad, a University of Arizona agricultural economist. This vulnerability creates myriad challenges for these businesses. The Bureau of Labor Statistics' list of the top twenty-five declining careers ranks farming and ranching at number one. Providing recreational activities and overnight stays is a way for families to stay on the land and push back against that depressing statistic.

"Small farms are under particular economic pressure to compete with commercial agriculture and are generally not able to produce the massive quantity of cheap fruits and vegetables grown on large industrial scales," says Tronstad. "So their challenge is to intentionally stay small and produce high quality, healthy organic options as an alternative."

Large or small, 97 percent of farms and ranches in Arizona and across the United States are also family farms, according to the United States Department of Agriculture (USDA), with 86 percent of U.S. agricultural products sold produced on these family farms or ranches. "When someone uses the word 'commercial' or 'industrial,' we assume it's non-family. In most cases, it is a family farm and, regardless of size, in America they are producing some of the safest, nutritious, and economical food around," says Julie Murphree, Arizona Farm Bureau.

Southern Arizona has more private land available compared to the other parts of the state, enabling more people to become farm owners or ranchers in the first place. Agri-tourism is a "relatively small but locally significant and expanding component of the economy" of Southern Arizona. Thirty years ago, few agricultural producers in Arizona were directly marketing their crops and value-added products to the general public. Today, visitors can buy farm products on-site in most Arizona counties and on many Native American reservations. "Much of this resurgent interest in farming and ranching is because families want to know where their food comes from," explains Murphree. "By 2050, almost seventy-five percent of the world's population will be urban. As a result, we have no contact with the source of our food other than eating it."

The effort to expand Arizona's agri-tourism began in the late 1980s, when Julie Leones, a University of Arizona College of Agriculture and Life Sciences (CALS) extension economist in the Department of Agricultural and Resource Economics, began to quantify the economic impact of agri-tourism in Cochise County through a research survey. The results were a catalyst for developing an agri-tourism notebook used to this day. This resource offers Arizona growers valuable information on planning their agri-tourism enterprises.

In 1988, University of Arizona cooperative extension agent Deborah Young put together the first "Southeastern Arizona Fresh Farm Produce" brochure. This brochure and later coverage by *Sunset* magazine and Phoenix and Tucson–area media contributed to the growth of the region's agricultural tourism. Since then, the amount of vacationers from around the state and elsewhere staying on or visiting farms and ranches as part of their vacation experience has increased exponentially.

Facebook, Twitter and other social media sites also spread the word. "We are working like crazy to use social media to market these places," says Murphree. Through these social media channels, anywhere from twelve to fifteen thousand Arizona families are connecting with the Arizona Farm Bureau about Arizona farming and ranching. Local First Arizona is also leveraging local businesses and social media networks to let people know about all the "cool places" to eat and shop while you are on a farm stay, wine tour or ranch vacation.

Many smaller farms are in danger of being absorbed by larger entities, with nearly eighty thousand farmers and ranchers predicted to be either in a different profession or out of work by 2018. "Agri-tourism will hopefully fight back against that projected decline," says Scottie Jones, owner of Farm Stay U.S.

It may be working. According to the USDA, since 2003, there has been an uptick in the number of small farms in business. "Size of a farm can be misleading," adds Murphree as a caveat. "Although some might seem small in acreage, the unit cash value of their products might be very valuable."

Farming and ranching as industries have a ton of potential to grow and create real jobs for their communities. "Whether small, medium or large, our Arizona farms and ranches are uniquely positioned to educate the public," observes Murphree.

Chapter 2

Courting Relaxation

A Brief History of Guest Ranching

"A dude ranch is a place where we can live out some of our fantasies."
—Joel Bernstein

Author Oren Arnold theorizes that the man who started guest ranching in Arizona was Charles Poston, also known as the "father of Arizona." In the mid-1800s, Poston visited the ranching village of Tubac, just south of Tucson. "I never dreamed of such a charming country. I'm a lawyer out of Washington, D.C. and let me tell you this—people back there'd give anything just to spend one month out there," said Poston of the area. However, in 1854, without a railroad, Southern Arizona was still largely inaccessible to large amounts of tourists.

Poston established roots in Tubac and became many things to the town—mayor, confessor and chief counselor, among other titles, for a short time. He eventually moved back East but was unable to forget the region's haunting landscape. He returned "for a spell of rest on the front porch of a ranch house somewhere" and, in doing so, became the first steady guest and customer of the local ranchers. Poston eventually gave up traveling and adopted the "strange, dangerous, beautiful region for his own." Southern Arizona's Sierra Bonita ranch, once Arizona's largest ranch, provided hospitality, elegance and even "cuisine" to travelers and famous guests like Wyatt Earp and Doc Holliday during this time period.

For the most part, accommodations and amenities on Southern Arizona's earliest guest ranches were austere. They were first and foremost working cattle ranches. Guests were encouraged to participate in the work of the ranch, and "Western traditions" helped immerse guests in the culture of ranch life. Even then, these early guests tended to return year after year, as the experience of guest ranching had made its impact.

By the 1880s, the establishment of rail routes provided easier movement of people and goods through Southern Arizona, and cattle ranching would peak during this time. This peak brought with it overgrazing and drought, and ranchers took in guests to supplement their loss of income.

By the turn of the century, Southern Arizona's landscape and ranches came to symbolize an almost mythical "unspoiled frontier oasis," as well as the American ideals of a return to simplicity, freedom and independence in the public's imagination of the day. Over the next few decades, through tourism the region gradually came to epitomize the "Western experience" in the minds of many "civilized" Easterners seeking adventure and escape.

Interestingly, the ways in which travelers reached their ranch vacation destination over the years is a microcosm of the changes in transportation in Arizona and the nation as a whole. The first guest ranch vacationers arrived by horseback, then stagecoach, then by trains and finally automobiles.

World War I would close Europe as a vacation destination, causing would-be travelers in the United States to consider their domestic alternatives. The "golden age" of guest ranching began in the 1920s and would last throughout the 1950s. The city of Tucson alone boasted over one hundred guest ranches open for business during this time. Many of Arizona's ranches became the place to be for young men from wealthy Eastern families. Although these young men were called guests, they were expected to do their fair share of ranch work and to "become healthy, hearty men." In the 1930s, Joe Kennedy sent a young JFK out to a ranch in Benson, a small town in Cochise County, to "toughen him up." Later, JFK invited the rancher John Speidon to his presidential inauguration. Speidon offered the president a job on the ranch should he ever need one. Kennedy politely declined.

Pop culture contributed to the growth in popularity of guest ranching and its accompanying image creation. Owen Wister's book *The Virginian* (also a long running TV show and movie), writers such as Zane Grey and Mary O'Hara, painters Frederic Remington and Charles M. Russell and the motion pictures of the 1930s all created a national public interest in the Western United States. A stay on a guest ranch promised an appealing blend of sophistication and leisure juxtaposed enticingly with the stark beauty of Southern Arizona's natural environment and appealing wintertime climate. "The romance was created early on, and

The Dude Wrangler
by Gail Gardner

I'll tell you of a sad, sad story,
Of how a cowboy fell from grace,
Now really this is something awful,
There never was so sad a case.
One time I had myself a pardner,
I never knowed one half so good;
We throwed our outfits in together,
And lived the way that cowboys should.
He savvied all about wild cattle,
And he was handy with a rope,
For a gentle-well reined pony,
Just give me one that he had broke.
He never owned no clothes but Levis,
He wore them until they was slick,
And he never wore no great big Stetson,
'Cause where we rode the brush was thick.
He never had no time for women,
So bashful and so shy was he,
Besides he knowed that they was poison,
And so he always let them be.
Well he went to work on distant ranges;
I did not see him for a year.
But then I had no cause to worry,
For I knowed that some day he'd appear.
One day I rode in from the mountains,
A-feelin' good and steppin' light,
For I had just sold all my yearlin's,
And the price was out of sight.
But soon I seen a sight so awful,
It caused my joy to fade away;
It filled my very soul with sorrow
I never will forgit that day.
For down the street there come a-walkin'
My oldtime pardner as of yore,
Although I know you will not believe me,
Let me tell you what he wore.

He had his boots outside his britches;
They was made of leather green and red,
His shirt was of a dozen colors,
Loud enough to wake the dead.
Around his neck he had a 'kerchief,
Knotted through a silver ring;
I swear to Gawd he had a wrist-watch,
Who ever heard of such a thing.
Sez I "Old scout now what's the trouble?
"You must have et some loco weed,
"If you will tell me how to help you,
"I'll git you anything you need."
Well he looked at me for half a minute,
And then he begin to bawl;
He sez, "Bear with me while I tell you
What made me take this awful fall."
"It was a woman from Chicago
"Who put the Injun sign on me;
"She told me that I was romantic,
"And just as handsome as could be."
Sez he, "I'm 'fraid that there ain't nothin'
"That you can do to save my hide,
"I'm wranglin' dudes instead of cattle,
"I'm what they call a first-class guide."
"Oh I saddles up their pump-tailed ponies,
"I fix their stirrups for them too,
"I boost them up into their saddles,
"They give me tips when I am through."
"It's just like horses eatin' loco,
"You can not quit it if you try,
"I'll go on wranglin' dudes forever,
"Until the day that I shall die."
So I drawed my gun and throwed it on him,
I had to turn my face away.
I shot him squarely through the middle,
And where he fell I left him lay.
I shorely hated for to do it,
For things that's done you cain't recall,
But when a cowboy turns dude wrangler,
He ain't no good no more at all.

people wanted a sample," says Marshall Trimble, author and Arizona's official state historian.

In the 1920s and 1930s, access to the region's guest ranches was primarily by railroad. The Santa Fe in Northern Arizona and the Southern Pacific in the south heavily promoted ranch tourism. Advertising brochures highlighting individual ranch room rates, reservation procedures, and the kind of activities guests might expect were available to passengers on board. An increasing number of tourists in the 1920s began to arrive in Arizona for their ranch vacation. At the same time, the region's established cattle ranches were experiencing financial troubles due to drought and falling cattle prices, so they again sought to supplement their incomes with tourist dollars.

Guest Ranches Today

Over the last few decades, grazing permits have been reduced for individual ranches in Arizona. Less grazing opportunity equals less livestock. Additionally, a combination of perceived health risks and increased competition from alternative meat products has caused the demand for beef to decline for years. Recreation, however, is an industry that continues to grow rapidly. Since 1970, Americans have steadily increased their spending on recreation and related goods and services regardless of the economy and personal income.

Recognizing this trend, many ranchers around the West now complement their existing agricultural operations by offering ranch tours, guided horseback riding, in some cases hunting and ranch stays with various amenities and activities. Operating a guest ranch can be a viable tool for enhancing a ranch's income, increasing cash flow and educating the public. "Some ranchers view it as their mission to educate the public on grazing and what ranches contribute. Getting people on their ranches is often the best way," observes Russell Tronstad at the University of Arizona Agricultural Extension. Public education may be the key to the survival and perpetuation of the ranching way of life. "Most people have little experience with agriculture, [and] when guests ride and see cattle grazing, they often comment on how happy the cows look," says Sarah King of Southern Arizona's Kings Anvil Ranch.

In the West, every ranch is a mixture of privately owned land with a large part leased from the state or federal government. Many ranchers are, by default, stewards of this public land. "These ranchers have a love for the industry and the land in which it takes place. Operating a dude ranch is a ton of work, and at times, the only payback they get is the privilege of living this lifestyle," says Colleen Hodson of the Dude Ranchers Association (DRA). As the governing body of the West's dude ranch industry, the Dude Ranchers Association's code of ethics, which all members must abide by, is evidence of this stewardship:

> *The Association promotes the western ranch vacation, while continuing to build a stronger working relationship with Federal and State land agencies in order to preserve and protect parks, forests, and wildlife. Members must adhere to all regulations, and provide information to clientele, concerning the responsible and sustainable use of resources. Members and the Association will work in cooperation with Federal and State land agencies to ensure the preservation of resources and habitat.*

"We have not deviated from these since 1926, so it is a very important part of who and what we are," says Colleen Hodson.

The term guest ranch replaced dude ranch in the early days and was meant to refer to a more upscale property. Literature from the DRA states that guest ranches can be broken down into the following three categories:

Working Dude Ranch

These are working cattle or sheep operations. Your horseback riding adventures will be determined by the ranch's livestock and the work related to them. Be prepared to experience these activities firsthand.

Dude Ranch

Horseback riding is central to these ranches. The cowboy in you will experience Western riding and a variety of outdoor activities.

Resort Dude Ranch

Horseback riding is featured, and these ranches offer an array of diverse activities and on-site facilities. These are apt to be the larger ranches.

Terminology aside, for most guests, weather will be the deciding factor in when to take a ranch vacation in Southern Arizona. Most ranches in Southern Arizona experience their best weather during the months of October through April or May. Some ranches, such as the Circle Z Ranch in Patagonia, only operate during this time.

Many guest ranches today have all the modern charms and amenities of a hotel or resort, such as gourmet meals, spas, massage services, heated pools and much more. "Some tourists still have the perception that you will be sleeping on old uncomfortable beds, eating beans and hot dogs and having to live without any amenities while staying at a ranch," says Hodson. The more "prototypically Western" activities such as horseback riding, rodeos, cattle drives, line dancing, cookouts and hay rides can be enjoyed as well.

Hodson recommends deciding what type of experience you are looking for in your prospective vacation as a first step. "If you are taking small children, make sure the ranch offers a kids program. If you would prefer to be around adults only, then you should be looking at adult-only ranches. If you want to be pampered a little more, a resort ranch might be your best bet," she says. "There is a perfect ranch out there for every taste. It is just a matter of finding the right one for you and your family." Regardless of which ranch vacation you choose, Oren Arnold writes, go there "determined to court relaxation."

Despite a high cost of doing business, guest ranches are continuing to grow their businesses as vacationers, like the early guest ranch patrons, are lured by the appeal of wide-open spaces and unique adventures. The experience of staying on a guest ranch offers guests many small but significant benefits. "One of the most common statements I get from ranch guests is 'I took the kids on a cruise a couple of years ago, and they remember very little about it. We still, after many years, are talking about their horse, their wrangler and what fun the entire family had at the ranch,'" recounts Hodson.

"Ranchers have great time-honored values," agrees Trimble, who has worked in various capacities at many guest ranches around the state. In contrast to a large hotel or resort, "you don't have all the distractors at

Lingo of the Land

brand—helps ranchers keep track of their cattle on the open ranges. Each ranch has its own unique brand symbol.

carrying capacity—the number of head of livestock that a ranch can sustain or support.

casita—guest house.

cowboy—from the Spanish word *vaquero* meaning mounted cattle workers. A cowboy's job is to work the cattle.

deeded land—private land.

dudes—guests.

hacienda—main house or headquarters of a ranch, a term used in the Southwest.

land grant—large tracts of land granted to settlers by the King of Spain during the Southwest's Spanish conquest.

ranch—classically a place where sheep and cattle are raised.

retreat—a vacation in which scheduled activities are workshop-based and/or of your own making.

spread—total amount of land that encompasses a ranch.

trail riding—horseback riding along trails, often with a guide, out in the surrounding ranch lands.

wrangler—goes out and gathers horses for cowboys to then select which horses would be used for the work of the day on the ranch. On a guest ranch, "dude wranglers" are in charge of all the horse activities for guests—gathering, saddling up and guided rides.

night, there's not a whole lot happening. You are dealing with the business of what a cowboy does, sitting around the fire, and playing cards. If you do those same things that at a resort, it's just not the same experience."

"A guest ranch vacation has the ability to bring a family back together where everyone sits down to dinner at the dinner table and talks about their day and their experiences," says Hodson. "Young people are not playing video games but actually getting out into the great outdoors. Parents are able to put the electronics down and really talk to their kids. There simply is not another vacation option out there that will bring a family together like time spent on a dude ranch."

Chapter 3

HACIENDA CORONA DE GUEVAVI

Nogales

THE SOUTHWEST'S PLYMOUTH ROCK

The Hacienda Corona de Guevavi is a romantic place. Wind chimes; décor that's a blend of Mexican, African and Southwestern influences; warm breezes; lush garden flora; a terrace overlooking gentle ranchlands; secret, tucked-away courtyards; and the distinct feeling of almost being in another country all add to that romance.

This is also very much someone's home—family pictures in the living room, chickens near the restored barn and a grandchild that offers to help with dinner all give the place its homey feel amid the elegance. The art interests of the hosts are obvious right away, from their collection of books to carefully chosen antiques. There is plenty of elbowroom here, however, both literally and figuratively. Even with quite a few guests staying at the hacienda during its high season, it is very easy to find a quiet space to be alone for a while.

The Hacienda Corona de Guevavi ranch house sits overlooking the east banks of the Santa Cruz River, near the Arizona/Mexico border town of Nogales. Although county records show that the adobe house was constructed in 1939, the history of the ranch stretches back to 1691 when Father Eusebio Kino, a Jesuit priest, arrived in what was a Pima Indian village with another Jesuit, Juan Maria Salvatierra. Kino and Salvatierra were the first Europeans to visit Guevavi, which in the Pima language means "big spring." They established a mission of the same name soon after. "I like to call this area the Plymouth Rock of the southwest," says Wendy Stover, the hacienda's owner. "This is where all the settlement started."

As a frontier settlement of New Spain, Guevavi was a home for 1,400 Pima Indians and a center for Spanish soldiers, miners, ranchers and Jesuit missionaries. Cattle and sheep were introduced at Guevavi in 1728 by Juan Bautista de Anza Sr., the leader of the army that escorted Father Kino to Guevavi, making this the oldest cattle ranch in the state.

In 1767, the Jesuits were expelled from Mexico by the Spanish government, and the mission at Guevavi was abandoned. The only Spanish Jesuit mission in the United States, the old mission ruins are currently owned by Tumacacori National Park, which runs tours to the mission ruins every month. Guests can also take a fifteen-minute walk from the Hacienda Corona to see these ruins.

Prudencia Acevedo Benedict and her husband homesteaded the ranch in 1915. The Benedict House, their former home, can be seen just after driving through the main ranch gate and is now owned by the city of Nogales. Mysterious buried coins from the Hapsburg dynasty were discovered just outside the building predating Kino's arrival.

In 1956, Ralph and Marjorie Wingfield purchased the ranch. The Wingfields expanded the ranch to slightly less than a 100,000-acre spread, making this one of the largest working ranches in Arizona in its day. Wingfield and John Wayne developed a close friendship during the filming of Wayne's classic film *Red River* near Sonoita when Wingfield lent some of his cattle for use in the film. Wayne, who had been coming to this part of Southern Arizona to film movies since childhood, became a frequent guest at the ranch.

Kai Eisline, who lived in the Benedict House (his father was the foreman of the ranch during Wingfield's ownership), remembered as a young boy trying to swagger through the swinging kitchen doors just like his idol John Wayne one day, not realizing that Wayne himself was sitting right there. He was so surprised that the door hit him unexpectedly, and he fell down in a less-than-manly fashion. Wayne would tease Kai thereafter, saying, "Well, have you learned to walk through a door yet?" Wingfield and Wayne even sailed around the world together on the actor's yacht. Shortly before his death from lung cancer in 1979, Wayne gave his cowboy hat (a cowboy's way of saying goodbye) to the rancher.

Over the 1950s and 1960s, the Hacienda hosted the Hollywood elite following Wayne's lead—Ronald Reagan, Elizabeth Taylor, Walter Cronkite, Stephanie Powers and many others lined the ranch's gate in their black limousines.

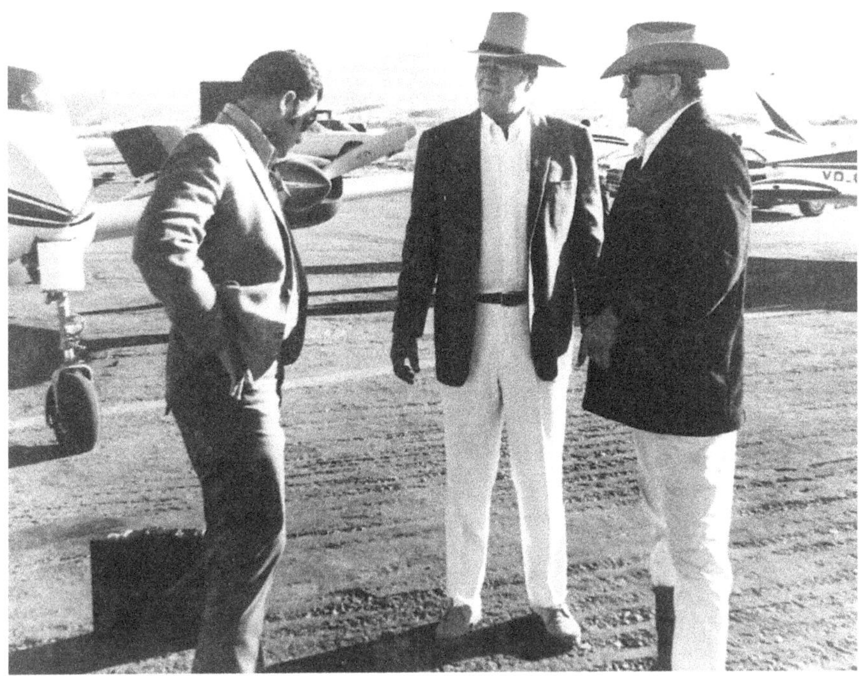

Former hacienda owner Ralph Wingfield embodied the hard-driving rancher that John Wayne often portrayed in his films. *Courtesy of the Hacienda Corona de Guevavi.*

Wingfield owned the ranch until his death in 2001 at the age of ninety-one. In 2002, Phil and Wendy Stover, from Connecticut, purchased the ranch. Wendy was a movie industry administrator and Coca-Cola consultant who had been looking for a way to get out of New York and the corporate world. She chanced upon a listing for Wingfield's twenty-four-acre ranch and hopped the first flight to Arizona.

Wendy and her husband grew up and met in Tucson and had always longed to return to the area. Many fortuitous "coincidences" made the property stand out—the ranch is located just off Ruby Road in Nogales, and Wendy had a dog named Ruby and an uncle with a ranch off Ruby Road, where Wendy spent time after her father died. "I found the area so healing," she remembers. A friend of Wendy's couldn't help remarking, "If that's not cosmic, I don't know what is."

Being interested in "cattle and cattle alone," Wingfield had left the place in a state of disrepair. But Wendy saw potential for a bed-and-breakfast. "I had always wanted to fix up an old farm house." The hacienda's mix of acreage for the Stover's horses, murals and Hollywood background made it

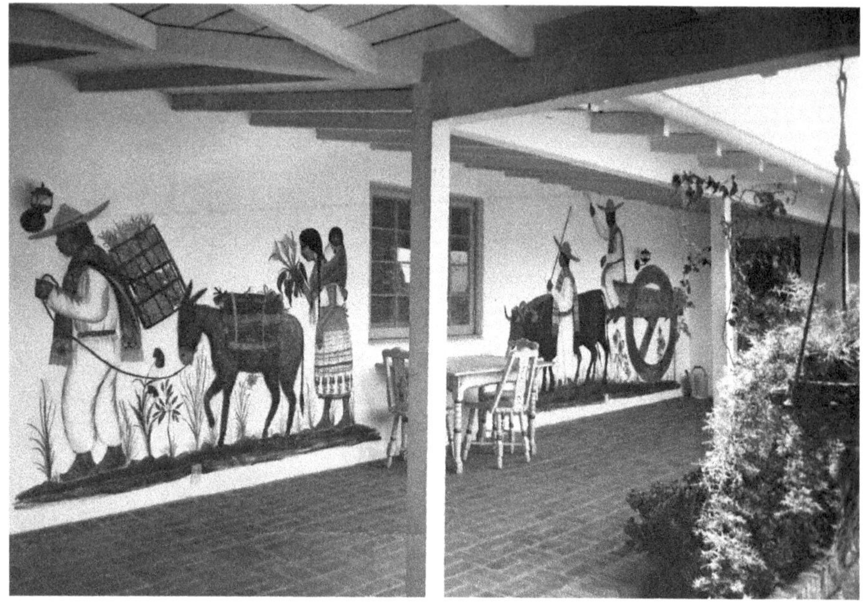

Famous Mexican bullfighter, folk artist and muralist Salvador Corona painted the hacienda's courtyard walls with these enchanting peasant scenes in the 1940s and 1950s. *Courtesy of the Hacienda Corona de Guevavi.*

seem a perfect fit. They soon sold their Connecticut home and restaurant and were living at the ranch full-time with a massive, life-changing endeavor on their hands.

The Stovers renamed the bed-and-breakfast after the famous Mexican bullfighter and folk artist Salvador Corona. In the 1940s and '50s, Corona painted its courtyard walls with scenes of Mexican peasants. The only murals known to have been painted by Corona for a ranch property, they were commissioned by the ranch's previous owner, Dines Nelson. Corona, best known for his colonial subjects and hand-painted furniture, featured indigenous subjects in these murals, perhaps inspired by the history of the old mission. Corona had a number of prominent clients during his lifetime, including Mrs. Gary Cooper; Mrs. John Dewey, wife of the famous American philosopher; and even President Franklin Delano Roosevelt.

The hacienda has plenty of room to board horses alongside the Stover's thoroughbreds and Arabians and Teddy Wingfield's painted pony, a gift to the Stovers. Horseback riding is available, and bird watching at the hacienda offers the chance to see many species of rare birds indigenous to Southern

Arizona. Dances in their adobe barn and tamale-making parties every spring are can't-miss events.

Over wine and hors d'oeuvres that evening, the conversation of guests centers on riding, horses, travel and the hacienda's incredible legacy. "What's amazing to me is how many people drive up to the ranch and say they used to live and work here," says Wendy, while serving our delectable refreshments. "The Wingfields' niece is even planning her wedding out here soon."

The corporate rat race a distant memory, the Stovers have gladly hosted fundraisers for local charities benefitting various social causes in the area. Their proudest accomplishment, however, is having preserved and continued the heritage of one of Arizona's historical treasures. "It was a lot of work but worth every hour," says Wendy.

Contact Information: 348 South River Road HC2 Box 96, Nogales, AZ 85621, Tel: (520) 287-6503, E-mail: stover@haciendacorona.com, Website: www.haciendacorona.com

Accommodations: Open year-round, the Hacienda Corona de Guevavi's comfortable yet luxurious guest rooms are each uniquely furnished to reflect the heritage of its past. Separate from the main ranch house are three private casitas designed with families, pets and extended stays in mind.

Chapter 4

Rancho De La Osa

Sasabe

"The mountains are not a barrier but an event."—David Burnham

To Melt the Heart of a Conquistador

Pristine desert landscapes protectively engulf this ranch with an almost intimidating pedigree. Settled in the late 1600s to early 1700s by Father Eusebio Kino and his Spanish Jesuits followers, Rancho De La Osa is still very remote, sixty-five miles away from the nearest traffic light.

Kino erected a mission outpost to aid the weary "padres and their explorers" who stopped to have horses shod, rest and get supplies before moving on toward California. In 1812, La Osa became part of the Mexican Ortiz Brothers' land grant from the King of Spain, which consisted of a total of 1.5 million acres stretching from Altar, Sonora, to Florence, Arizona. The Gadsden Purchase of 1854 settled the border dispute between Mexico and the United States, and the ranch then fell within U.S. boundaries.

After the Ortiz brothers, the first known owner and the builder of La Osa's hacienda or main house was Colonel William Sturges of Chicago, Illinois, who purchased just under two hundred acres sometime before 1891. Sturges lived on the ranch full-time and later married Spanish Dona Leonora, who operated a trading post east of the Baboquivari Mountains.

In 1894, Sturges sold the spread to James Finley, who increased the acreage to 320 acres. After his death in 1899, property ownership changed again, and the ranch was sold to Lyman Wakefield and Edward L. Vail,

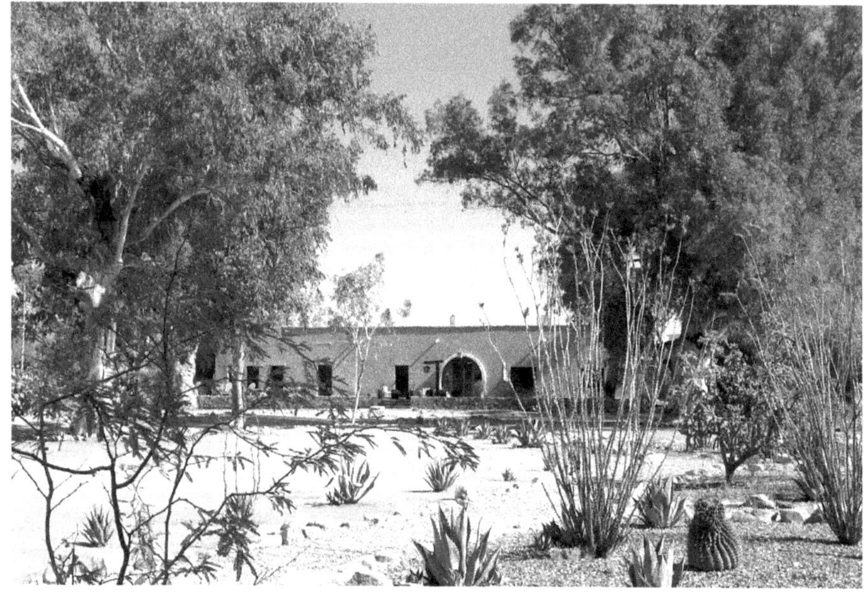

The centuries-old Rancho De La Osa is reputed to contain the second-oldest building in Arizona. *Courtesy of Rancho De La Osa Guest Ranch.*

owners of the nearby Empire Ranch. The ranch today encompasses 250 deeded acres and 300 acres of state-leased land.

The origin of the ranch's name, translated as "ranch of the she-bear," is not fully known. Some say a Mexican cowboy killed a bear and her cub near the ranch, and so the Spanish name for the bear also became the name of the ranch. A De La Osa family was prominent along the border from the time of the ranch's earliest settlement, though there is no solid connection between the ranch and that family.

"In the early 1920s, Louisa Wade Wetherill and her husband, John (who discovered the Mesa Verde [ruins] in Colorado with his brother) came to Sasabe from Kayente, Arizona, looking for a lost tribe of Navajos. The Wetherills owned a summer guest ranch in Kayente and decided La Osa would make a wonderful winter guest ranch," explains Veronica Schultz, the ranch's current owner, describing the early days of the ranch's transformation into a guest ranch. The ranch opened in 1924 with a rodeo and dance.

Over the years, Rancho De La Osa's architecture has been featured in numerous publications across the world. Tucsonian R.M. Pacheco built the east and west guest wings using adobe bricks and is credited with bringing to the ranch a "return to the Mexican pueblo (of New Mexico) traditions

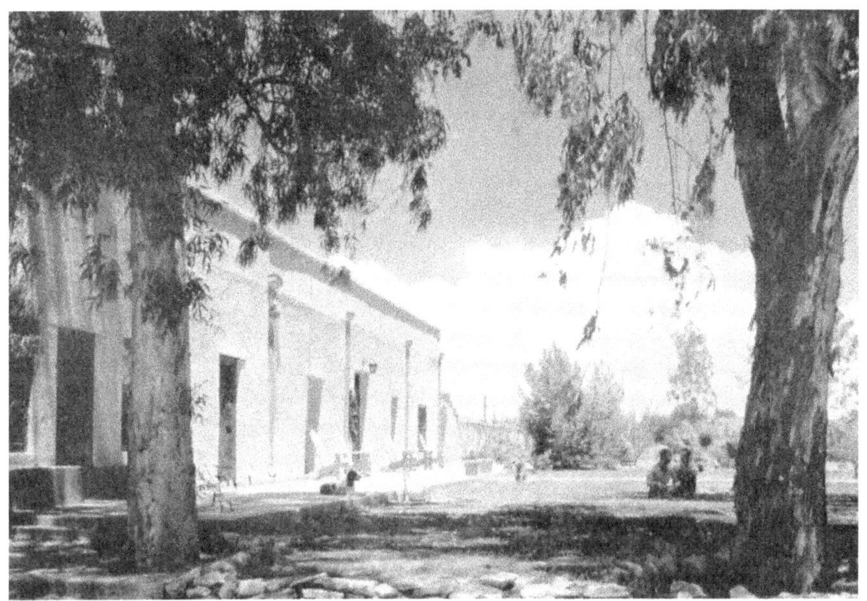

Opened as a guest ranch on Thanksgiving Day 1924, the color and romance of the old Spanish Southwest still lingers at Rancho De La Osa. *Courtesy of Rancho De La Osa Guest Ranch.*

during the 1930s and 1940s which had become romantic and high style," writes Janet Stewart in *Arizona Ranch Houses: Southern Territorial Styles.*

In 1933, Richard Jenkins from Pomoukey, Maryland, purchased the ranch, and it flourished during this time. Jenkins's grandfather, Dr. Mudd, had set James Wilkes Booth's leg, which was broken when he leapt to the stage after assassinating President Lincoln. Jenkins and his sister Nellie also ran La Osa as a guest ranch. During their years of ownership, they expanded its quarters by constructing the southeast guest cottage. It was in this guest cottage that the Marshall Plan was written in 1947. The cottage was built for Marshall Plan coauthor Secretary Clayton's wartime use when the ranch served as a secluded retreat for Washington guests, as well as the Carnegie Mellons, Joan Crawford, Cesar Romero, President Lyndon Johnson, John Wayne, Tom Mix and many English guests. Robert Sherwood and author Margaret Mitchell were also among the visitors to have frequented the ranch during this time.

Although La Osa itself is no longer a working cattle ranch, during the early 1900s through the early 1950s, over five thousand head of cattle passed through the ranch on their way from Mexico to Tucson to be slaughtered. The Schultzes' current neighbors to the northwest graze cattle on their ranch.

In 1996, Veronica and Richard Schultz bought the ranch and continue to manage all aspects of the operation. From the moment they saw the ranch, they realized it would be a lifelong project. The Schultzes were inspired by the ranch's tremendous history "to continue a lifestyle that is dying. Guest ranches are remote, and fewer and fewer exist every year," says Veronica.

When arriving at the front gate of the ranch, the property's color and blissful silence strike you instantly. Ocotillo fences function as protectors from the monsoon weather. "We enable guests to live without TVs" and constant noise from cell phones. They can, in a sense, "come and listen to the silence nature provides."

The inside is just as welcoming, and its eclectic beauty leaves me momentarily speechless. The main hacienda consists of a large dining room, which seats thirty; a small dining room, which also has a sitting room; and a commercial kitchen where the ranch's home-cooked southwestern meals are prepared by a chef. Paintings of the southwest, many the works of artists who visited the ranch in earlier years, decorate the walls. A vintage black-and-white photograph of four little boys wearing cowboy hats and standing with their bicycles in the middle of a dirt road catches my eye. The old Jesuit mission, the oldest building on the ranch and said to be the second-oldest building in Arizona, dating to the late 1600s to early 1700s, has since been converted into the "Cantina," a comfortable bar. A library filled with western novels, along with many books donated by guests, and a living room with a piano and large open fireplace greets guests as they gather before meals.

A small, gated historic cemetery that dates back to the 1700s can be reached by taking a short walk up a hill overlooking the ranch and its gentle countryside—perhaps the ranch's most meaningful feature. Many previous owners, as well as people who have worked on the ranch over the years, have been laid to rest here. The cemetery's unmarked graves are said to be victims of the 1918 Spanish influenza.

Daytime activities at the ranch include horseback riding along its many trails and through the nearby Buenos Aires National Wildlife Refuge, guided hikes, nature walks, road biking and bird watching. A hot tub, pool and massage services are available for relaxation and enjoyment. Cowgirl Camp, Couples Roundup, cattle roundups, cooking lessons, wine-tasting dinners and informal Western dance instruction are examples of the ranch's "themed weekends."

The Schultzes felt a sense of responsibility for and instant connection with La Osa's land and its wildlife. But the greatest challenge the ranch faces today is not managing its fragile environment, but rather it is "getting and

Rancho De La Osa is noted for its superb southwestern cuisine with a distinctive blend of Spanish, Mexican, Native American and Southwestern ingredients perfected over the years. The ranch has been featured in five cookbooks. "The Arizona Centennial Commission chose one of our signature dishes Pinon Crusted Chicken with Cherry Chipotle Sauce as the official entree for the Arizona Centennial Menu," adds Veronica. Enjoy this recipe courtesy of the ranch owners.

Pork Tenderloin with Maple Syrup-Balsamic Sauce

4 (eight-ounce) pork tenderloins

Marinade
2 cups dark beer
½ cup soy sauce
2 tablespoons chopped cilantro
2 tablespoons diced jalapeno
2 tablespoons fresh lime juice
2 tablespoons brown sugar
Mix above ingredients. Marinate the pork tenderloins for at least 2 hours and no more than 4 hours.

Sauce
2 cups chicken broth
½ cup maple syrup
¼ cup balsamic vinegar
¼ cup tomato ketchup

To make the sauce, bring the chicken broth to a boil and reduce by ½. Add the maple syrup, balsamic vinegar and tomato ketchup, then bring to a boil and again reduce in volume by ½. Set aside in a warm place.

maintaining good ranch hands." Working on a ranch is "more of a way of life" than a nine-to-five job.

Guests are drawn each year to the ranch's nature, the beauty of the mountains, peace and quiet, cuisine, historical relevance and "the antiquity of our buildings." Tradition is big here. Every night, La Osa still rings their one-hundred-year-old bell to summon guests to meals served at a table made for and delivered to the ranch in 1932. This table was made from one pine

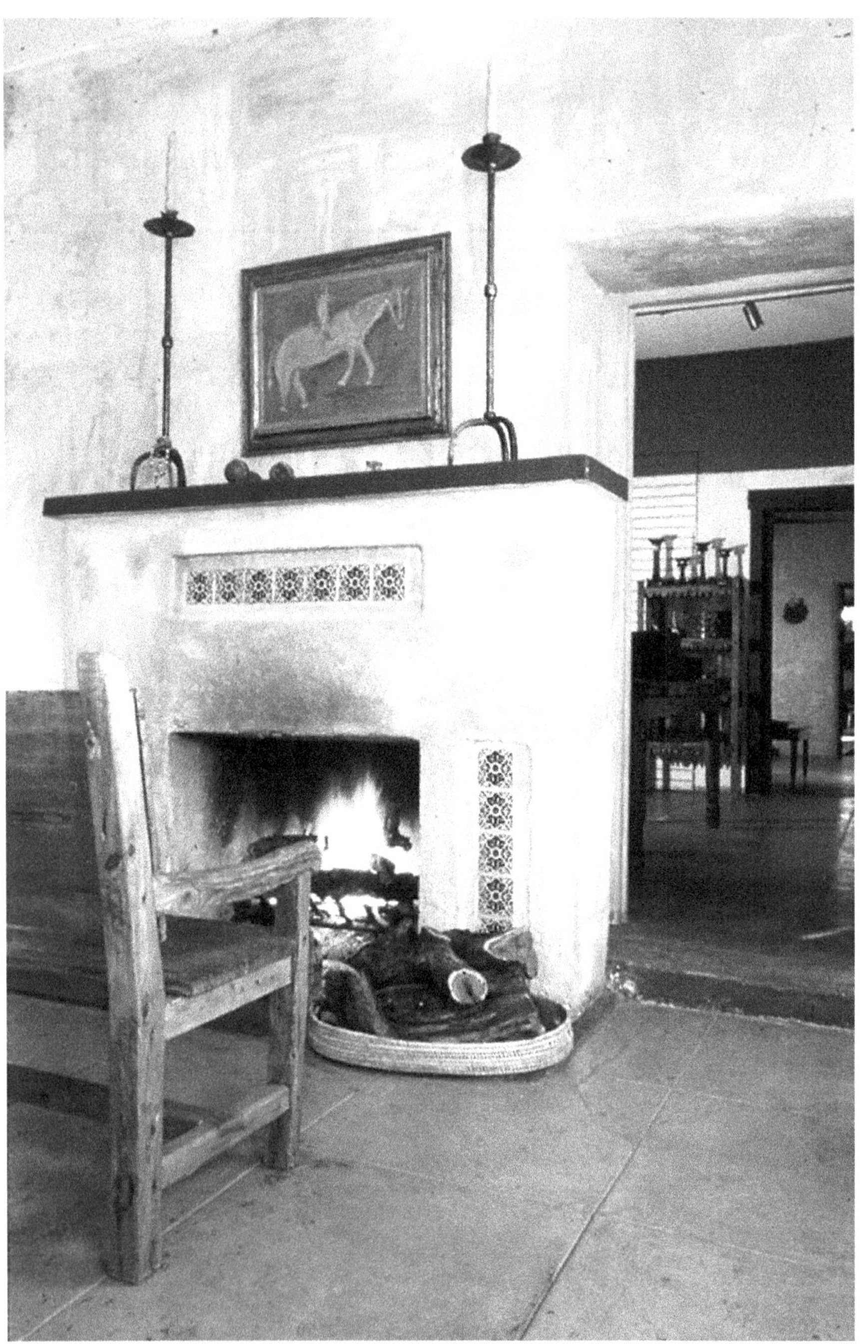

Enjoy a blazing fire and the sound of silence in Rancho De La Osa's hacienda salon. *Courtesy of Rancho De La Osa Guest Ranch.*

tree from Mount Lemmon in Tucson. "As it is one giant table, everyone sits together. You can meet the other guests and have great conversations. You never feel lonely, even if you are traveling alone. Our guests get to know all of our staff. Friendships are started here between guests and staff, something that would never happen on a large property," Veronica explains. "We encourage everyone to be sociable with each other."

One of Veronica's favorite ranch memories happened years ago during one such sociable cookout. "I offered a male guest a Portobello, and he said: 'Did you just call me a portly fellow?' My other favorite was when I asked one of my kitchen helpers to put the corn in the oven…the Spanish word for corn is 'elote,' and I told him to put the 'tecolate' in the oven, which means 'owl.' We have a pair of great horned owls who have been here since long before I arrived. It was quite funny," she laughs.

Since coming to live and work on the ranch, the Schultzes have learned to appreciate every day. The Tohono O'odham believe the 7,700-foot peak of Baboquivari Mountain, seen in the distance, is the "Home of the God of Creation" and the "Center of Mother Earth." "Even though we are remote, I realize as I gaze at our blue skies on a daily basis that this truly is a little piece of heaven," says Veronica.

The Tohono O'odham seem to have it right.

Contact Information: One La Osa Ranch Road, Sasabe, AZ 85633, Tel: (520) 823-4257, E-mail: osagal@aol.com, Website: www.ranchodelaosa.com

Accommodations: Each of Rancho De La Osa's eighteen adobe guest rooms is furnished with Mexican antiques and vintage furniture. A private porch opens to mountain views, and most have a wood-burning fireplace.

Chapter 5

Cold Creek Ranch

Clifton

"In a place of all-encompassing silence,
any sound is something to be noted and remembered."
—Sandra Day O'Connor

The Very Essence of Western Ranching

The drive from Tucson to Cold Creek Ranch is magical. Land and sky seem to merge into one. Scattered boulders, cacti, puffy white clouds, vibrant colors—nothing detracts from the power of this landscape.

A ten-thousand-acre working cattle ranch, Cold Creek Ranch is located twenty miles east of Clifton in Arizona's Mogollon Rim Country. The Schwennesen family has roots in Arizona since 1921, when Otto M. Schwennesen returned from World War I with lungs damaged from German gas attacks and was sent to the Veterans' Hospital at Fort Whipple (now Prescott) for treatment.

Cold Creek produces 100 percent grass-fed beef. Although the family property has been a ranch for many years, it has just been open to guests since 2011. The ranch is comprised of part state land, part National Forest and part privately owned. Originally established in the 1870s as part of the larger Davis Ranch, Cold Creek is the geographical name for the stream that carries two other creeks from their confluence to the Gila River. During these more lawless times, "you managed as much as you could manage and

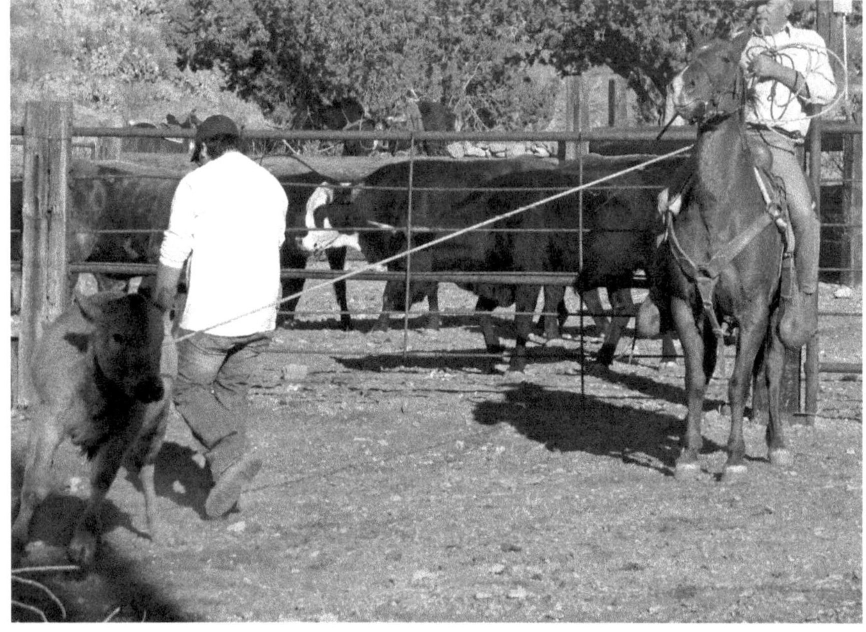

Originally established in the 1870s, Greenlee County's scenic and remote Cold Creek Ranch offers guests the chance to share in a rare lifestyle. *Courtesy of Cold Creek Ranch.*

afford to pay for," says ranch owner Eric Schwennesen. "People more or less claimed a part of the land, and hopefully no one challenged you. The land is too rough, wild and difficult for any other occupation."

Originally, ranching in the area was extensive, with few neighbors or competitors. Today, ranching in Greenlee County is "intensive and limited, with many regulation constraints," explains Eric. "Whole government bureaucracies have been created to oversee and micromanage ranching." Water is in short supply, and "everything else waits on that. If it doesn't rain, nothing happens," he says.

Greenlee County is perhaps the ultimate Southwestern historic ranching region, and little has changed over the years in this part of Arizona. It is comforting to be in a place where little changes. Supreme Court justice Sandra Day O'Connor grew up on the nearby Lazy B Ranch. This is also said to be the birthplace of Apache chief Geronimo and the heart of the cattle and mining region since the 1880s. However, this way of life is "dying out fast, our communities along with them. Mining and ranching are about all we have in Clifton," says Eric. "Clifton and the other scattered towns in the area were built to supply the growing, dynamic needs of livestock and mining industries over a wide region; today, the only thriving industry is that of government regulation," he says. "Our town depends on our success as much as we do."

The Schwennesens bought Cold Creek ranch in 2006 after their former ranch, Double Check in Pinal County was being consumed by commercial speculation. "It seemed like hundreds of houses were being built by the minute. Developers were always looking at our water to supply these new developments," remembers Eric. "You could see the writing on the wall," he remembers sadly.

The boom in Pinal County's housing construction affected every significant piece of property, most of which were eventually subdivided. Noise pollution and crowding—open space was getting harder to find—also contributed to the family's decision to move. "People that go to guest ranches for a vacation are looking for the same things we were looking for," says Eric. "We couldn't do anything else once we started managing the land; we didn't want to. After you've done this, there isn't anything else you can do that has any significance. You are on the interface between heaven and earth."

Eric and his wife, Jean, are no armchair academic ranchers, although between them they hold two bachelors' and three master's degrees. "We learn from other people in the business, from the very scattered system of pastoralists around the world." It is essential that you actually work as a rancher. "I won't push anything on people that I can't do myself," he says. "You don't need a college degree to manage the land," he adds. "I still can't find anyone from Western universities that can give a scientifically valid definition of overgrazing."

Too far to have reasonable access to electrical lines, Cold Creek Ranch is "off the grid," meaning they generate their own power with solar panels and batteries to store energy. In the winter, they heat the house with firewood; in the summer, "you live with it." Eric has no regrets though. "It's a good life, close to the land...There's that vital connection in managing the land well," he says.

The ranch directly markets meat locally and doesn't envision feeding into the megastore pipeline. "There is absolutely no reason to ship halfway across the country. There are local producers who need the business and support in every state, including Alaska. We are not food extremists, however, just common sense folk."

One inspiration for starting Cold Creek Ranch was to correct the negative public perception of ranchers. "Too few people understand land husbandry or the difficulty of it; this is a way to let them experience what it is like.

Ranching has endured unjustified bad press for fifty years—not many people ranch in the first place, and a diminishing few understand what it takes. Living in a city separates you from that," he says. "There are organizations that have grown very wealthy in the last thirty years by fabricating a 'ranching is bad' philosophy, and now that they have become wealthy, many of them have been buying ranches!" Reflecting on the current state of ranching in this country, Eric first recommends "getting rid of the enormous multilayered federal bureaucracies" and "allow[ing] land management to take place…If you had the same situation in a city today, a schoolteacher would have to submit a lesson plan to the scrutiny of departments of Pencils, Homework, Classroom Size, Building Violations, Ink Toxicology, Discrimination, Transportation, Risk Assessment, Nutrition—and any one of those departments finding fault with your lesson plan would force you to start over," is his hilarious parallel. "It's all control and no management."

The sincerity of Eric's love of the land and passion for this work rings loud and clear. A day's work on the ranch in which guests might participate includes any number of tasks—riding, feeding and shoeing horses, calling and moving cattle, roping and branding (in season) or clearing brush. Well known in ranching, making prickly pear syrup is a unique experience for guests. Jean's recipe uses agave syrup instead of sugar, which "apparently has quite notable anti-diabetic qualities." Day trips from the ranch include hot springs, Apache sights and world-famous Tombstone. The ranch's "Weather Guarantee" states that any day with .35 inches of rain will be free for guests.

What brings guests to Cold Creek ultimately is its "scenery, remoteness and legitimacy; this is not a movie set," says Eric. "The real deal is what we offer, actual work, a part in the management process, a chance to see how it is done, a chance to learn and share in a rare lifestyle."

Contact Information: 160372 AZ Hwy 78 Clifton, AZ 85533, Tel: (520) 705-1525 E-mail: eschwennesen@gmail.com, Website: www.coldcreekranch.org

Accommodations: Cold Creek is open for guests from October 1 to April 30, with weekend and weeklong stays recommended. In addition to private rooms in the main ranch house, a small bunkhouse is available for rental on a self-catering basis. Three meals a day are served during ranch stays and can be customized to follow the South Beach diet principles.

Chapter 6

CIRCLE Z RANCH

Patagonia

"Climb aboard a horse and examine the traces of those who have gone before."
—Fred R. Egloff

WHERE PRESERVATION IS A KEYSTONE THOUGHT

Upon arrival at the ranch through an iron gate with its charismatic livestock brand "Circle Z," you are first struck by geography and landscape. The ranch sits at an altitude of four thousand feet in a high-desert environment on almost seven thousand acres. Giant sycamore, ash and cottonwood trees provide a cool and shade often not associated with Southern Arizona. Sonoita Creek, one of the few year-round, free-running, surface-water courses in the state, flows through the ranch. Across the creek, Sanford Butte, also known as the "Circle Z Mountain," serves as a landmark for riders and hikers seeking to return to the ranch from a distance. Hikes and trail rides leading out from the ranch provide extraordinary views of the Santa Rita and Patagonia Mountains, the Santa Cruz Valley and Mexico.

Guest ranches have a certain warm, coming home quality to them, and it is very much this way at the Circle Z. You feel taken care of immediately. Homemade cookies and fresh lemonade, roaring fires and a genuine staff definitely contribute to this feeling.

The Circle Z's amenities include a heated pool, tennis court, cantina and game room. Trail riding, cowboy poetry and cookouts are some of the

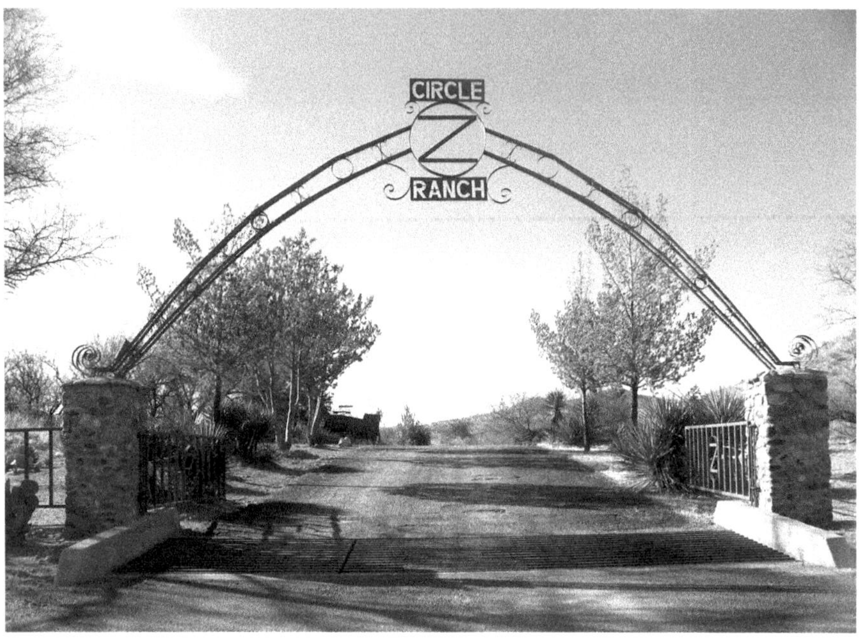

"Guests all have ideas of what the ranch will be like. They come along the two-lane blacktop and turn at our gate. They fall into their dreams, and the ranch fills them up, the ranch takes care of them," observes the Circle Z's ranch manager and former guest Jock Soper.

ranch's many activities to enjoy. However, the ranch is extremely "peaceful, with activities that are not lockstep...Staff go with the flow of what guests want to do," says long-time guest Fran Karanovich, who has been coming to the ranch for many years as an "annual rest haven." She says, "I thought spending a thousand dollars to ride horses in the winter was crazy, but now I come every year." The ranch has a definite effect on its guests and is beloved by all who visit. One guest from back East even decided to relocate to the Southwest after staying at the Circle Z.

After a delicious breakfast of (huge!) blueberry pancakes in the Circle Z's quiet, intimate dining room, Jock Soper, the ranch's current manager and a former guest himself, shows me the garden and greenhouse that provide fresh produce for the ranch and where bees are raised for their honey. The Circle Z is "moving toward becoming a self-sustaining ranch, as it had to be in the Territory days," he explains, through "harvesting sunshine to replace fossil fuel, growing food—fruits, vegetables, beef and pigs," says Jock. "The goal of ranching, with or without guests, is to achieve harmony with the cycle of things—nature, culture and economy."

The Circle Z is Arizona's oldest continuously operating guest ranch. In 1874, Denton Gregory Sanford, a native of New York, became the ranch's original owner and settler. Sanford homesteaded on Sonoita Creek across from the present-day ranch complex. Considered one of the most elegant haciendas in the Arizona Territory during its time, the adobe ruins of the homestead are still visible today.

The heirs of Don Leon Herraras, of Tubac, took over much of the ranch from Sanford under an 1825 Spanish land grant. In 1923, Sanford's daughter Bertha Miller sold what was left of the ranch to Lee Zinsmeister of Pennsylvania. "The Zinsmeisters had searched for quite a while to find the perfect spot for a guest ranch. They wanted it to be spectacular, but also convenient so that guests could easily ride off in any direction to explore. They found exactly the spot, south of Patagonia, and that is today's Circle Z Ranch," comments the ranch's current owner, Diana Nash.

Zinsmeister built the ranch up to be one of the oldest and finest guest ranches in Arizona during the country's golden age of guest ranching. Guest facilities were opened in 1926. Some guests even reached the ranch via their own private railway cars—railway was the most efficient way to reach the ranch in those days. Polo was the popular sport of the times, and the Circle Z even had its own polo field. The ranch team played against military teams from Fort Huachuca and club teams from both sides of the border in Nogales. The Circle Z's annual Fourth of July picnic became one of Santa Cruz County's biggest attractions. Over the years, Hollywood has used the ranch and its environs many times. Portions of classic movies *Red River*, *Broken Lance* with Spencer Tracy, *McLintock* and *Oklahoma* have been filmed here or nearby.

In the 1930s, Circle Z became the home of El Sultan, the Spanish stallion owned by the ranch. The horse was a gift from the Royal Spanish house, before its abdication. He was the only one of his breed (Cartuja Spanish sire) in the United States and the breed most likely used by the Spanish Conquistadors. Mrs. Zinsmeister, known to be an "entrancing hostess" and elegant woman, would dress in fancy silver-trimmed vaquero (cowboy) regalia to entertain guests with trick riding on El Sultan. To this day, the Circle Z continues to raise and train many fine horses, and most are also bred on the ranch. It takes six to seven years of training for a young horse to get into the "guest string," explains Jock. Their horses are cared for year-round and throughout their lives. When age overtakes them, they live out their lives in comfort in a big ranch pasture.

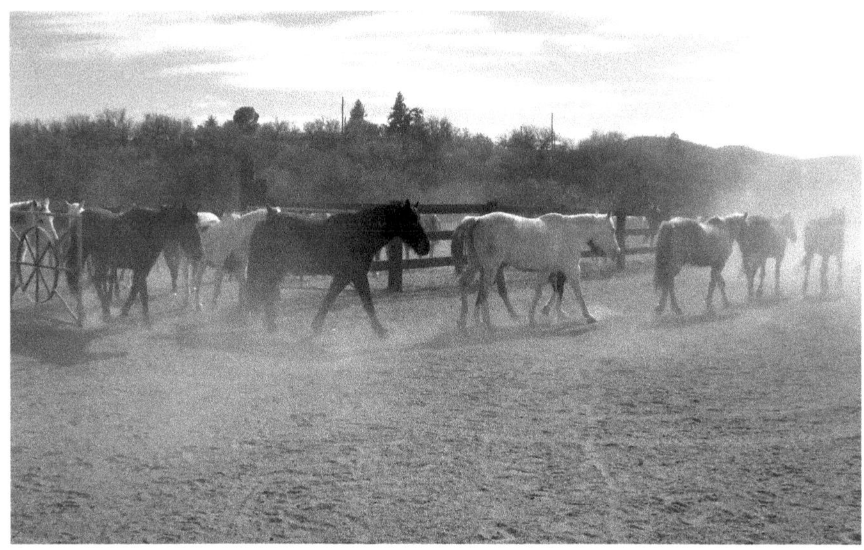

The Circle Z Ranch is celebrated for its large herd of fine horses that are raised and trained on the ranch over the course of their lives. Pictured here, roughly one hundred horses excitedly gallop to their evening pasture after a day of trail riding.

World War II slowed the resort and travel business all over the state, and the Circle Z changed hands several times. In 1949, Fred Fendig, from Chicago, purchased the ranch. For the next twenty-five years, he was its owner and manager. When Fendig sold the ranch in 1978, rumors circulated that it would become a tennis ranch or large development. Then Mr. and Mrs. Preston Nash of Novelty, Ohio, purchased the spread and changed its destiny. It was their intention to preserve and operate the ranch in keeping with its original traditions as a "warm, old-time, family-style ranch with an emphasis on good riding, good food and congenial guests."

Lucia Nash's many visits to the ranch as a child dating back to the Zinsmeisters' ownership influenced the family's concern for and desire to save the ranch. "Lucia Nash first came to the Circle Z when she was eight years old, in 1937. She came by train with her family from Ohio and spent many childhood days roaming the hills and canyons on horseback," says Diana Nash, Lucia's granddaughter. "Grammy purchased the ranch in 1974 to preserve the land but also to preserve her memories of the place. We have tried to keep the ranch the way it was—simple, but exquisite in little ways. We want to preserve the sense of what Lucia saw as a little girl so that the ranch can be enjoyed by many generations to come."

At its core, Diana feels that "preservation is a keystone thought at the Circle Z…Preservation of the landscape, preservation of the unique biological diversity of the region and preservation of a way of life that was rugged, careful, generous and mindful of the gifts given by the place itself."

A morning ride at the ranch takes you through the evidence of this preservation and concern. Riding trails run through The Nature Conservancy's lush Patagonia-Sonoita Creek Preserve. The Nash family donated a portion of their property to the preserve, and in exchange, guests use these trails for recreation. Famous for its bird migration and rare bird nesting, the wildlife and nature viewing during the ride is superb—a massive 155-year-old sycamore tree, tadpoles swimming in the creek, an owl, a blue heron. "Deer like to lie down in the tall grass, so you want to keep hold of your horse when you're in that environment," instructs Jenny, the ranch manager's wife, during the ride.

After roughly two hours riding on the trail, our group arrives in Patagonia for lunch at the local saloon. Arrival in town is very exciting. I truly feel I've accomplished something with my day and have certainly created a lasting memory. It is impossible to fully capture in words just how much fun I had riding that day.

It's no wonder that Diana Nash says that "the Circle Z makes memories effortlessly." One of her fondest memories comes from the year her son, Preston, was two years old. "Our foreman bought Chiquita, a tiny horse just for Preston. Grammy loved to lead Chiquita with Preston in the saddle and take them on walks along Sonoita Creek or out to the old hermit's homestead on the ranch. That was when I began to understand the family connection to the place, how important this ranch would be to all of us."

The love of the ranch and connection to its place that guests communicate is also striking. This particular group is already planning their next trip out here. "It's the totality of the experience, the fellowship between staff, management and guests. Every year, I can't wait to come back. Every time you come, it's a new experience," says devoted guest Fran, on our way back to the ranch.

Location: P.O. Box 194, Patagonia, AZ 85624, Tel: (888) 854-2525, E-mail: info@circlez.com, Website: www.circlez.com

Accommodations: The Circle Z's adobe cottages, furnished with area rugs and antique Monterey pine chests, are available for singles and families.

Chapter 7

Tanque Verde Ranch

Tucson

The Evolution of Tucson's Oldest Business

Secluded but not isolated, Tucson's Tanque Verde Ranch is located in the foothills of the Rincon Mountains and bordered by Saguaro National Park and the Coronado National Forest. The drive to the ranch is lined with smaller private ranches and open, little-developed land along East Speedway Boulevard, eye-opening for this Tucson resident who always associated this particular road with fast food, indistinctive shopping complexes and suburban sprawl. Tanque Verde appears at the very end of the road as the grandest ranch of all.

The country smell of horses and cattle permeates the air. Tanque Verde excels as a vacation destination while continuing the real work of a cattle ranch. The Cote family still grazes over six hundred head of cattle on 640 acres and a sixty-thousand-acre Forest Service lease.

Considered the oldest business in the greater Tucson area, Tanque Verde Ranch's history stretches back to the late 1800s, and it is one of the oldest, most prominent ranches in the state. Former Texas ranger William S. Oury was the ranch's first owner. Known for his combat skills, Oury was the leader of the infamous 1871 massacre of approximately one hundred Aravaipa Apache women and children at Camp Grant and would eventually become Tucson's first "Anglo" mayor.

In 1870, twenty-year-old Emilio Carrillo established his cattle operation at Tanque Verde. For several years prior, Carillo had worked a homestead

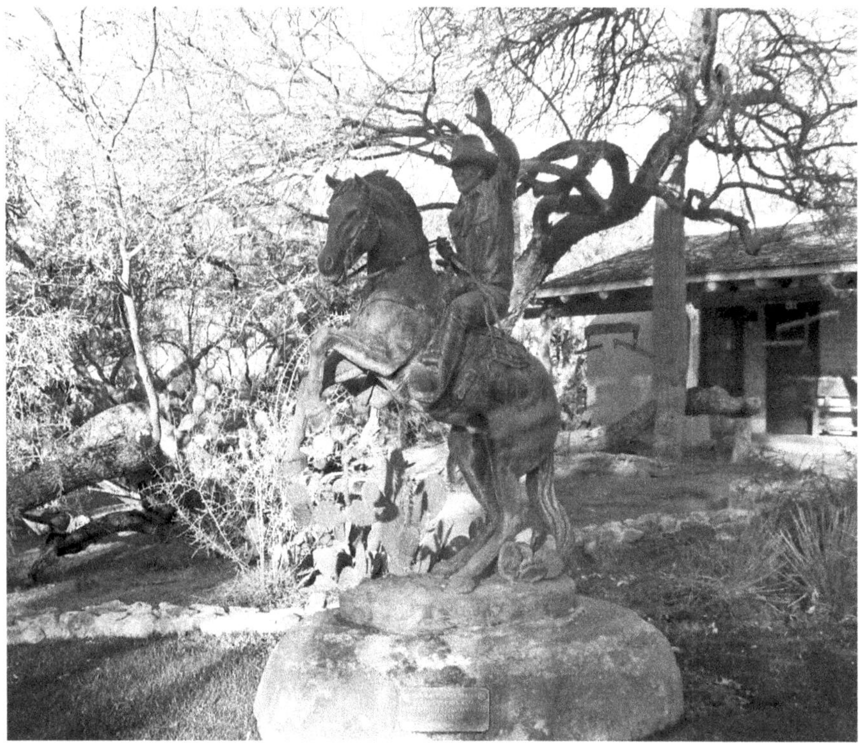

Tucson's Tanque Verde Ranch has been in the Cote family since 1957. A statue of Brownie Cote on horseback stands in front of the ranch's historic dining room, commemorating this incredible family legacy.

in Cochise County. With those profits, he began buying property around Tanque Verde. The Sonoran custom of planting barley on the earthen roof of his adobe house as insulation against the intense summer heat inspired Carillo to name his ranch Cebadilla or "Little Barley." Carillo's cattle brand became known throughout the southwest. Over time, his ranch prospered, and the value of Carillo's property in Tucson eventually totaled over sixty thousand dollars.

In 1904, rumors that Carillo had hidden a gold fortune somewhere on ranch premises caught the attention and imagination of local bandits. Emilio was ambushed while he was out riding with his son Rafael. Although he somehow managed to keep his hiding place (under the hearth of the kitchen stove) a secret, Emilio is said to never have gotten over his near murder. In 1908, he died. For the next twenty years, Rafael ran the ranch.

In 1928, Texas cattlemen James Converse and Alan Gray bought the property from Rafael. Converse loved to hunt and found ample opportunities in the mountain ranges surrounding Tucson. The two began inviting friends from back East to participate in cattle roundups and other ranch activities. During this time, Converse's friends suggested that he start taking on "paying guests," and soon, guest ranching at Tanque Verde was born. Jim had never liked the name "Cebadilla," so David Evans, a friend of his who ran an exclusive boys school, suggested the name "Tanque Verde" after the region's deep pools of water standing in the washes as a replacement.

Despite the peaceful name of the ranch, Tanque Verde has had its fair share of conflict and turmoil. One fateful night, Jim was drinking and started to quarrel with a friend. To frighten him, he shot up in the air but aimed fatally wrong. The man died, and Jim went to jail.

In 1957, Brownie and Judy Cote from Minnesota purchased 480 acres from Converse that included the Tanque Verde Ranch buildings. It was their forward-thinking ownership that allowed Tanque Verde to flourish and eventually create the modern ranch getaway that guests experience today. The guest ranch operation is now owned by Brownie's son, Bob Cote. "My father was looking for a property that was adjacent to public land to guarantee open space for riding and hiking and with vistas that did not overlook a subdivision," he says. "We have managed to maintain our historical flavor while adapting to the modern needs of our guests. You can't remain static and be in business."

Bob lives on the ranch himself and still cooks outdoor breakfasts for guests twice a week. "Living on the ranch is like welcoming friends to your home every day. Some guests we've known for years."

Today, over half of Tanque Verde's guests are return visitors and have been coming to the ranch for many years. One particular family has been coming to the ranch for over forty years just to play cards. Tanque Verde Ranch is perhaps best described as "a very large home where you entertain."

Original building materials of adobe brick, native stone fireplaces and desert plant wood for ceilings give the ranch its historic authenticity amid modern, resort-like facilities. Each guest casita has a porch facing the open, green, gorgeous Sonoran desert landscape. Taking an evening stroll around the pathways connecting various ranch buildings is like walking through a lush Sonoran garden wonderland.

With a herd of 180 horses, Tanque Verde Ranch is one of the largest guest ranches in the country. Trail riding, team penning (a timed competition between cattle and guests), family scavenger hunts, horseback riding lessons and customizable horsemanship courses immerse guests in the frontier spirit of the West.

This is a place to be active and outdoors. With an extensive amount of activities from which to choose, Tanque Verde can cater to the interest of each family member in your party—yoga, tennis, art, hiking, cooking, science and swimming, and the list goes on. "People are always surprised that we are not just a riding ranch," says Bob.

Tanque Verde's guest rates include three daily gourmet meals in their Southwestern-themed dining room, all trail riding and horseback lessons, fishing, water aerobics, mountain biking, art classes, tennis clinics and lessons, a children's program, outdoor BBQ cookouts and a breakfast ride up into the mountains. Arizona resident specials are available for non-overnight guests.

Perhaps their most unique feature, though, is a naturalist program with a nature museum highlighting the flora and fauna of the Sonoran desert. "We have always believed very strongly that the Sonoran desert is one of our treasured assets. Survival in the desert with its extremes in temperature,

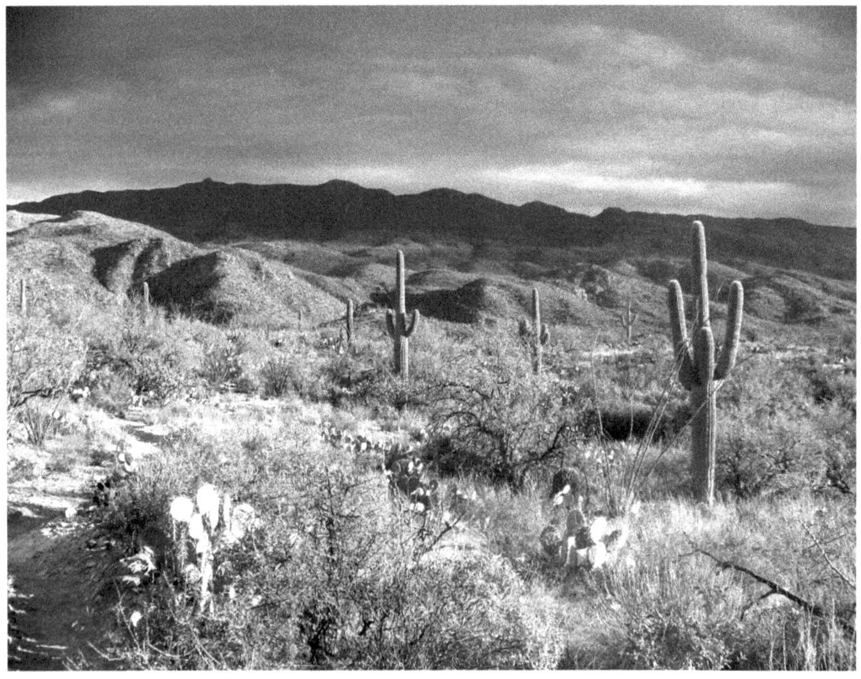

Tanque Verde's extensive daily naturalist program takes guests on nature hikes and interpretive walks through the lush Sonoran desert ecosystem.

drought and poor soil conditions has many unique solutions for both plants and animals. We have always tried to educate and stimulate our guests, and the Sonoran desert gives us an ideal springboard," comments Bob. Guests can participate in the naturalist program's daily nature hikes and interpretive walks. "On one walk, guests even saw a mountain lion sitting up in a tree, just watching all that's below. He looked curious, not afraid," remembers Marcia, Tanque Verde's resident naturalist.

One of the easier loop walks takes you past the ranch's fishing pond, through the desert landscape and finally up to Emilio Carrillo's ghostly old stone homestead that overlooks the sprawling newer ranch buildings below. Watching guests receive tennis instruction, recite wedding vows and relax by the pool's waterfall, it is clear that Tanque Verde has come a long way from its days as a frontier ranch fortress defending itself against Apache raids in a wild, lawless Southern Arizona.

Contact Information: 14301 East Speedway Tucson, AZ 85748, Tel: (800) 234-3833, (520) 296-6275, E-mail: dude@tvgr.com, Website: www.tanqueverderanch.com

Accommodations: A variety of luxurious lodging options complement swimming pools, hot tubs, tennis, volleyball and basketball courts, live animal exhibits and the ranch's La Sonoran spa.

Chapter 8

Historic C.O.D. Ranch

Oracle

Hospitality Returns to Oracle

Sturdy oak trees, stunningly beautiful boulders, blue and purple vistas of six thousand acres of the Santa Catalina Mountains and the Coronado National Forest unfold as you drive down a long, dirt road on to the Historic C.O.D. Ranch property. The C.O.D. letters stand for "Come on Down." Mountain lions and bears frequent the ranch's surrounding thirty-two acres located just outside of Oracle.

In the late 1880s, Frank Daily became the first homesteader to use this land. Daily and his thirteen children grazed large herds of cattle over thousands of acres. In 1927, the ranch was sold to Bill Huggett, who expanded the operation to include even more cattle and staff. After Mr. Huggett's death in the late 1930s, his wife, Elna, continued to run the ranch for ten years before selling. The majority of the ranch was split and sold to various people in the 1940s. Some of it was donated to the parks service and is now part of the Oracle State Park. The Huggetts' daughter Wilma lived in the "Round House" (a house that sits above the main part of the ranch) until her death in the 1990s.

In 1996, gregarious poet and Renaissance man Steve Malkin found the Historic C.O.D. Ranch in a very dilapidated state. Originally an East Coaster, Steve purchased, restored and refurbished the ranch with the intention of "bringing hospitality back to Oracle." It would eventually

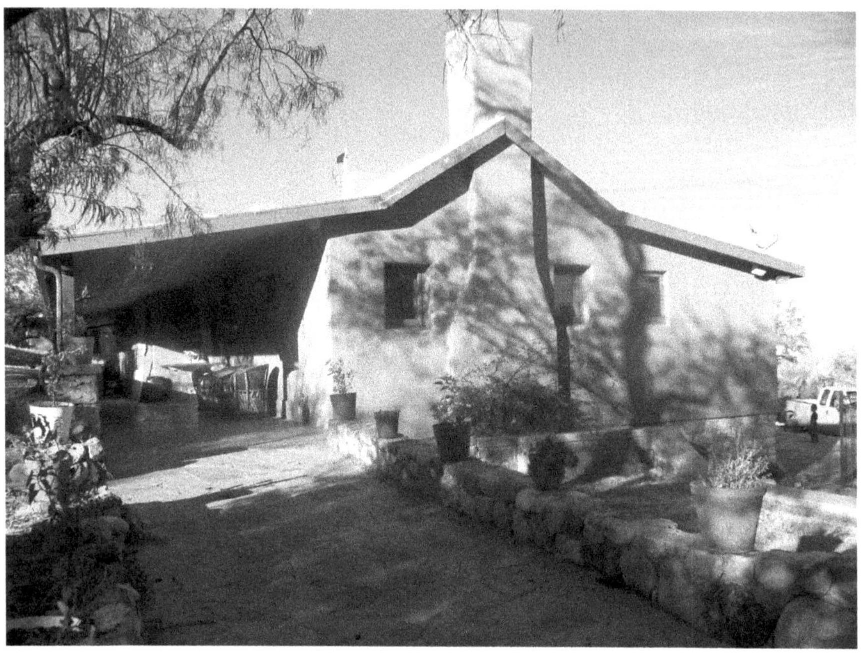

Poet Steve Malkin founded the present-day Historic C.O.D. Ranch as a retreat and guest ranch in 1997. Built in the 1920s, the ranch's main house is an excellent example of the architecture of its day.

become a premiere Southern Arizona retreat and special events center. In a short amount of time, Steve developed business acumen out of necessity to make the ranch the vibrant place he knew it could be.

In the 1970s, Steve had been one of Wilma Huggett's carpenters at the ranch. While he was living at the nearby Rancho Linda Vista Community for the Arts, Steve learned restoration, and he says that "a lot of my skills came from Linda Vista." He then traveled a great deal and restored large, historic buildings around the country. During his travels, Steve became president of the Galveston Foundation and started a poetry guild in Texas, among his many accomplishments.

When Linda Vista founder Charles Littler died, Malkin came back to Tucson (via Russia) to read a poem at his friend's funeral and to be with Littler's family. One day, Charles's wife, Pat, took him with her to the C.O.D. because, he explains, she was "becoming disillusioned with life at Linda Vista" and was looking for a place to live with her son.

In the 1980s, Biosphere 2 had bought the ranch from Wilma and had gutted its natural and built environment, according to Steve. "Wilma was going to take it back from Biosphere." At the time, Steve had "no intention" of moving back to Arizona, but the ranch seemed to be "calling out for help." A couple who wanted to tear the ranch down made their own financial offer to Wilma. Steve offered full price with the intention of restoring the ranch from the ground up.

Pat moved on to the C.O.D. Ranch with her son, and Steve got to work on his "hideout," as he describes it. "I didn't have a particular vision at the time. I just wanted to get the place fixed up. My commitment to restoration kept me going." In three years, the present day C.O.D. Ranch was born. Having spent time on guest ranches as a student at the University of Arizona, Steve wanted the ranch to "feel like the old days in Arizona."

Steve's friend, a midwife, thought the ranch's beauty and energy would make it a great place for a conference. After that initial success, the ranch soon became a retreat space for all kinds of groups and families. "The midwives gave birth to the ranch," Steve says fondly.

The C.O.D. also functioned briefly as a dude ranch with twenty-five horses and access to twenty-six thousand acres of Coronado National Forest trails. In 2001, the economy crashed, and a choice between keeping the horses or the ranch had to be made. Just in the nick of time, Native American groups saved this "cowboy" ranch with their added patronage.

Steve's weakness is restoring old buildings, but ironically, the "ranch restores people…In a way, we all need a little restoration," he says.

"The ranch is a return to simplicity." Through his travels, Steve has become more interested in "goodness than greatness," as he puts it. "I try to have the ranch be a place where everyone is treated with kindness. I wanted to create a place that's not local or in time. Guests can just be," he says. "It hasn't been easy, but I've been blessed. It hasn't been profitable, but it's been a calling, an honest effort."

Wilma ran the ranch very differently, he remembers. "She was more of a 'crack the whip' type. But deep down she had a big heart." One of Steve's favorite "Wilma stories" shows these dual sides of her personality:

> *Once, a guest hiked to the Round House with his family. Wilma, who was living there at the time, saw them coming and shot over their heads. They ran back down the path completely freaked out. But you couldn't tell Wilma anything. So I called the "Council of the Grandmothers" group and asked them to come to the ranch and have a kind of "powwow" with Wilma. The next morning, Wilma said, "Please don't call the grandmothers, I'm sorry."" I called the meeting off.*

The C.O.D. doesn't advertise. Steve wants people to discover the ranch on their own. The lasting impact that the ranch has made as a retreat center is obvious. Tailor-made vacation and retreat packages are crafted and adjusted to suit a variety of budgets and needs—spiritual and artistic retreats, business meetings, family reunions and individual vacations. "Guests can be alone to work on whatever they are working on," says ranch manager Stacy Raneri. "Personal growth, health and fitness and spirituality of all kinds."

Since the main ranch house, bunkhouse and barn were built in the 1920s, guests can see examples of ranch architecture of that period. As the ranch's popularity grew over the years, several casitas have been added to accommodate guests. "The idea is that the ranch is your own," says Stacy. A small but caring staff check in to make sure that guests have what they need. "Since we specialize in catering to one group at a time, we serve meals based on the group's agenda," she says. A world-class chef who has cooked in some of the best restaurants in the West creates healthy, satisfying meals freshly cooked with no processed foods. The ranch has even published their own cookbook.

The C.O.D.'s spiritual retreats are a "natural extension" of the original intention of the guest ranch as a place where people could gather in creativity and self-development. "Guests report taking so much away from the ranch, but actually we feel they leave just as much of themselves," says Stacy. Stone altars, paper lanterns, meditation circles and myriad works of original art are testament to this.

Minted Melon Soup

Courtesy of the Historic C.O.D. Ranch:

1 ripe cantaloupe
5–6 radishes
1 bunch fresh mint

Peel, seed and dice the cantaloupe.
Wash and stem the radishes.
In a blender or food processor puree the melon and radish. Add a few mint leaves at a time to taste. Chill, serve and enjoy.

With their wilderness locale, ranch management strives to be eco-friendly as much as possible. From the beginning of Steve's ownership, the ranch has made improvements in their environmental friendliness every year. Everything is recycled, and the ranch's original windmill towering above pumps water for the whole ranch. The surrounding Coronado National Forest has miles of trails for mountain biking, walking, birding and communing with nature. "Guests comment most on the peacefulness of the ranch and that they feel taken care of while they are here. We hear from almost every group that when they return that it is like a homecoming," says Stacy.

Events have happened at crossroad times that have fortuitously enabled the ranch to keep going. Steve has found that the C.O.D. has been "a lesson in listening to the earth in a down-to-earth sense. If you keep your feet on the ground, more spiritual things can happen."

Contact Information: P.O. Box 24137 C.O.D. Ranch Road, Oracle, AZ 85623, office: (520) 896-2883, mobile: (520) 850-2452, E-mail: staff@codranch.com, Website: www.codranch.com

Accommodations: The C.O.D.'s Round House and entire ranch of twenty-two separate rooms are available for personal, spiritual and company retreats or vacation rentals. All-inclusive rates include accommodations, three meals (in conjunction with a larger group) and refreshments. The Round House comes complete with a private kitchen, porch with mountain views and original artwork. The main ranch house features a large living room with a fireplace, fully equipped kitchen, breakfast/dining sunroom and a private bedroom. Visit off-season during June, July and August to enjoy the area's comparatively cooler summer temperatures and refreshing private pool. Stay tuned for the return of horseback riding at the ranch.

Chapter 9

Rancho Linda Vista

Oracle

Arizona's First Guest Ranch?

One of Arizona's earliest working guest ranches, Rancho Linda Vista is possibly even its very first. The ranch is located in small-town Oracle, with its "reputation as being eccentric, not following the normal social protocols," according to Chuck Sternberg, a local historian who has lived at the ranch for the past forty-four years and designed many houses in the area. "Oracle never became another Tubac or Sedona. There's a sense of authenticity and mix of people unlike other tourist towns. The people that live here wouldn't live anywhere else," says Chuck.

Now the ranch is an artist's community and home to roughly fourteen families, two horses, five cats, eight dogs and assorted wildlife. Rancho Linda Vista is a "huge resource for Tucson as a show space for the visual arts. With a close-knit arts scene, it is also the rural arm of Tucson's arts scene," adds former resident Steve Malkin and current owner of the nearby Historic C.O.D. Ranch.

The ranch is listed on the National Register of Historic Places. The old stagecoach road, once the way guests reached the ranch from Tucson, is still visible. In 1882, pioneering rancher Henry Bockman raised his cattle and built a small adobe dwelling, as well as some corrals, on the property that would later become known as Rancho Linda Vista. Although there were other prominent owners after Brockman, it was George S. Wilson, however, who would have the most lasting effect on the ranch's history. Wilson was a

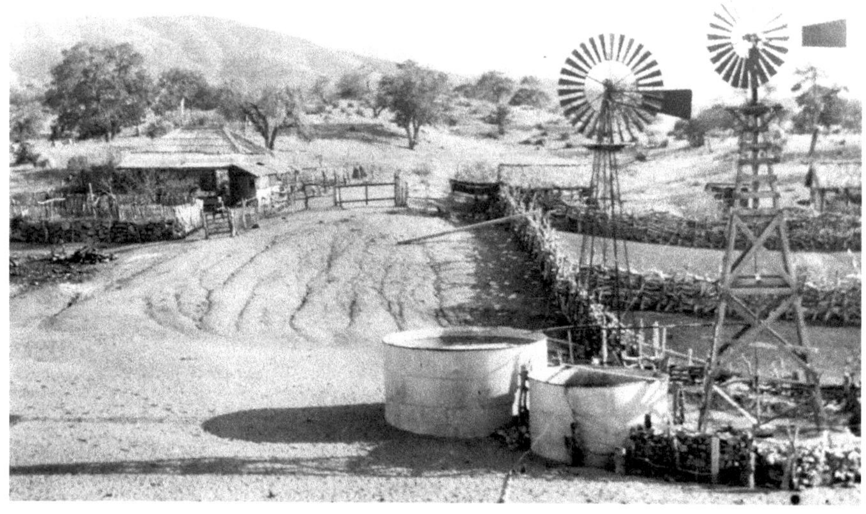

Rancho Linda Vista was founded by George Wilson in 1924 as one of Arizona's earliest working guest ranches. *Courtesy of the Oracle Historical Society.*

cattle rancher who became a "nationally-known pioneer of Arizona's guest ranching industry."

Born in Scranton, Pennsylvania, in 1906, Wilson was sent by his family to Oracle to recover from a baseball injury, one of many health-seekers who would come out to Southern Arizona. Wilson found Oracle's dry climate to be beneficial and soon established his cattle ranching business in the area at the age of twenty-three.

From 1912 onward, in addition to his main ranch near Oracle, Wilson acquired properties along the Cañada del Oro, a corridor for travel and European and American settlement from Tucson north. Wilson's spread grew to nearly 150,000 acres, with the carrying capacity of 1,500 head of cattle. He and his wife named their holdings Rancho Linda Vista, or "Beautiful View Ranch." In 1922, famous novelist Harold Bell Wright penned *The Mine with the Iron Door* while staying at Wilson's upper ranch.

After the region suffered a serious drought, the Wilsons, as well as other ranchers, lost a great deal of cattle. As a way to provide supplementary income, Wright, Angus Hibbard (superintendent of the American Telephone and Telegraph Company and one of Rancho Linda Vista's early guests) and other friends advised the Wilsons to start a guest ranch. In 1924, a skeptical

Wilson borrowed money to make the necessary additions and remodel the ranch to be able to accommodate guests.

Rancho Linda Vista became known as the premier guest ranch of its time. In 1925, Hollywood began making the movie version of Wright's book. Wright insisted that they film at Rancho Linda Vista, the real site that had inspired his book, inadvertently creating instant fame for the ranch. Undoubtedly impressed by the exoticism of the locale, the stars of the film returned the following year as the first of many movie stars and other well-known guests to become ranch patrons. Rita Hayworth, George Sanders, Gary Cooper and even Bugsy Siegel's girlfriend were among these Hollywood notables and some of the first "dudes" or guests of the ranch. Horseback riding was the most popular activity for these guests, with hunting and sightseeing expeditions, square dancing and fine food all contributing to the Western ambience of this destination.

During World War II, the tourist economy took a turn for the worse, and the ranch closed for a time. Rancho Linda Vista was run as a guest ranch until 1955. Wilson died two years later, and the property remained in the Wilson estate until 1958.

Then, Andy Warhol's Western movie spoof *Lonesome Cowboys* was filmed at Rancho Linda Vista in 1968. The film's subject matter caused something of a scandal in Oracle at the time, and the FBI even came out to investigate an alleged violation of interstate pornography laws. A few residents in town were interviewed by agents but had "seen nothing," laughs Chuck.

In that same year, a group of ten couples and families of art professors and graduate students from the University of Arizona, lead by artist Charles Littler, would collectively signed a mortgage on the property, creating the ranch's present-day incarnation as the Rancho Linda Vista Community of the Arts. The group's intention was to explore the possibility of communal living (it was the 1960s after all) based upon "artistic values."

At the time, the ranch was going into foreclosure, and the move seemed serendipitous as well as inexpensive. "There were already enough houses on the property (fifteen guest houses), so people could move in right away. We were never trying to be a commune though," comments Chuck. "We really had a good time those first few years—it was the late '60s, as you can imagine."

The community's first residents restored the adobe houses and installed electricity. The old barns were converted into studios and a public gallery. The entire ranch was transformed into the living-art gallery of today that blends effortlessly into the high-desert wilderness. Sculptor Joy Fox

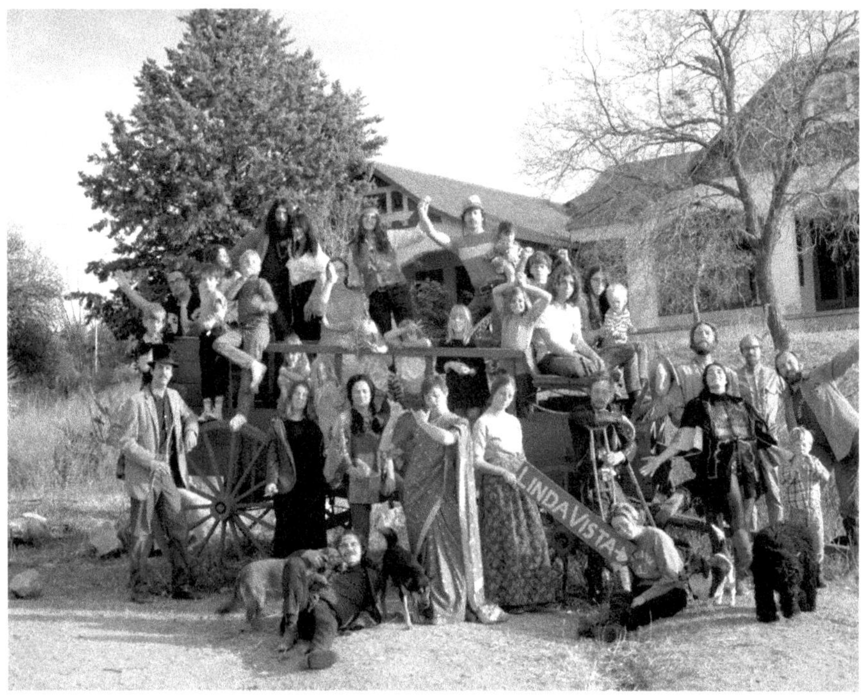

The original residents of the Rancho Linda Vista Community of the Arts in 1968. *Courtesy of Rancho Linda Vista.*

McGrew, who moved to Rancho Linda Vista in '68 with her children, then eighteen months and four years old, comments on those early years: "It was a wonderful place to raise a family and still is for my grandchildren. We are surrounded by nature, which is always inspirational for art. There is a constant flow of artists through our guest artist programs and gallery, which also enriches our lives and our own art. It has been an adventure working on our old adobe buildings, dating back to the 1880s, to preserve and protect the history of this special place and our original intention of creating an environment which is conducive to artistic endeavors."

Although he was an accomplished painter, sculptor and drafter, Littler called the community at Rancho Linda Vista his most significant work of art. He lived at the ranch until his death in 1992.

From its beginnings, Rancho Linda Vista Community for the Arts has provided a place for artists of all mediums to live and work. "Linda Vista is very hospitable to others. Whatever they've tried to create, it has endured," says Steve, one of the original groups of artists, after a recent trip back to the ranch. "Despite a lack of any endowment or grants, it has maintained its commitment to the arts and artists. That's something."

Residents are well-known specialists in their individual artistic fields of writing, painting, sculpting, cartography, building and shoemaking, to name a few disciplines. The ranch's "Visiting Artists Program" offers month-long residencies and retreats to artists from around the world. A wide network of artists and friends return to visit the ranch each year, and the ranch has sponsored hundreds of arts events.

"It's the evolution of a place. Linda Vista is like a human being—right now it's in middle age, but somehow we keep going," says Chuck. The vibe here is soft and serene, most likely what the earliest "dudes" experienced at this historic ranch in the foothills of the northern Catalina Mountains.

Contact Information: P.O. Box 160, Oracle, AZ 85623, E-mail: rancholindavista@gmail.com, Website: http://interstice.us

Accommodations: Rent a furnished cottage by the month to write, paint and work in the Rancho Linda Vista studio or "just be with the intention to be artistic."

Chapter 10

Triangle L Ranch Bed and Breakfast

Oracle

Natural and Created Places of Beauty

The smell of fresh basil, swiss chard, bok choy and arugula float through the air as I arrive at Oracle's Triangle L Ranch Bed and Breakfast, a guest ranch that is many things to many people. An outdoor art gallery, farm and community center, it's perhaps best described as a "multi-use facility," says local resident and composer Michael Carroll.

Evidence of the ranch's place in the community of Oracle is obvious right away, and even the infamously picky publication *Tucson Weekly* has given the place a good review. Every Saturday morning, the Triangle L hosts the Oracle Farmers' Market outside its barn-turned-gallery space. There are homemade scones, beef chili and locally roasted coffee. Eggs, tomatoes, organic honey and squash from the area's local farmers are also being sold this particular week.

"It's a gathering place for guests, a breakfast club and a small farmer's market," says Sharon Holnback, multi-media artist and the ranch's current owner, an extraordinary woman. The atmosphere is delightfully jovial. The gallery/barn is the venue for a variety of events, like concerts, lectures and workshops. Nearby is an outdoor kitchen featuring a clay oven built in one of these workshops.

At an elevation of 4,500 feet, the ranch encompasses fifty acres and more of a high-desert environment. Sharon purchased the ranch in 2001. "I was looking for some land outside Tucson that was unimproved," she

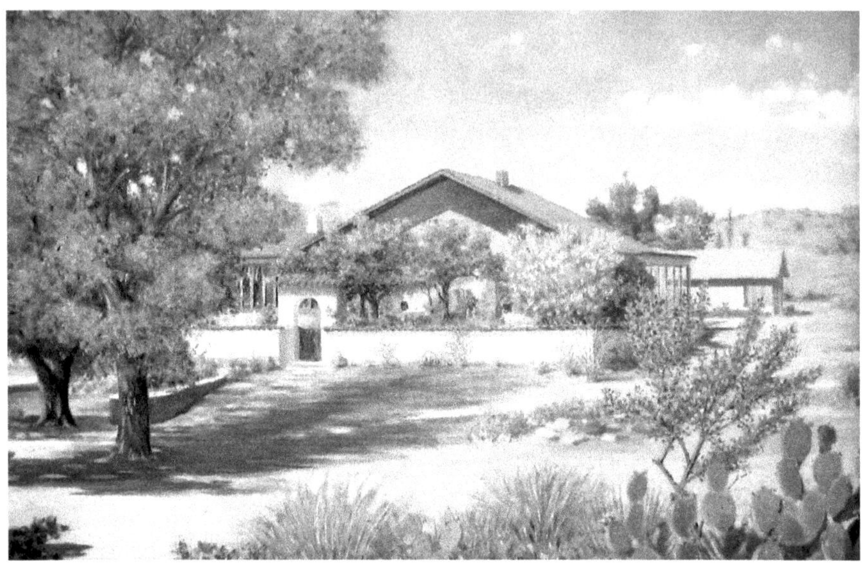

The rustic and real Triangle L Ranch Bed and Breakfast is a place to "relax and recharge," says owner Sharon Holnback. "You are surrounded by nature and history." *Courtesy of the Triangle L Ranch Bed and Breakfast.*

remembers. "I found the Triangle L with a colleague and saw potential. It seemed like such a treasure." Her "do-it yourself" retreats—family reunions, yoga and weddings—have been popular ever since. "People can utilize the facility for things they are interested in. The sky's the limit." Sharon's vision, in addition to the bed-and-breakfast aspect, is for the property to be a haven for artistic endeavors.

Once inhabited by Apaches, in the 1890s, the ranch was first homesteaded as the "Boot Ranch," after the ranch's leather-making activities. Then Westry Ladd, Bostonian architect and "gentleman" rancher, purchased the ranch. Buffalo Bill is said to have been a regular visitor at the ranch during Westry's ownership. Westry's wife, Laura, was an artist and painter and the heir to a locomotive company. "They had silver tea service, silver saddles and were very privileged," remarks Sharon. "According to the Westry family tree, they were most likely also cousins," she says. Westry willed the property to his brother William, who also operated it as a cattle ranch, although he left the real work to a ranch foreman and cowboys.

In 1924, William B. Trowbridge became the owner and turned the Triangle L into one of Southern Arizona's oldest guest or "dude" ranches. Trowbridge was a known philanderer as well as a philanthropist. Trowbridge

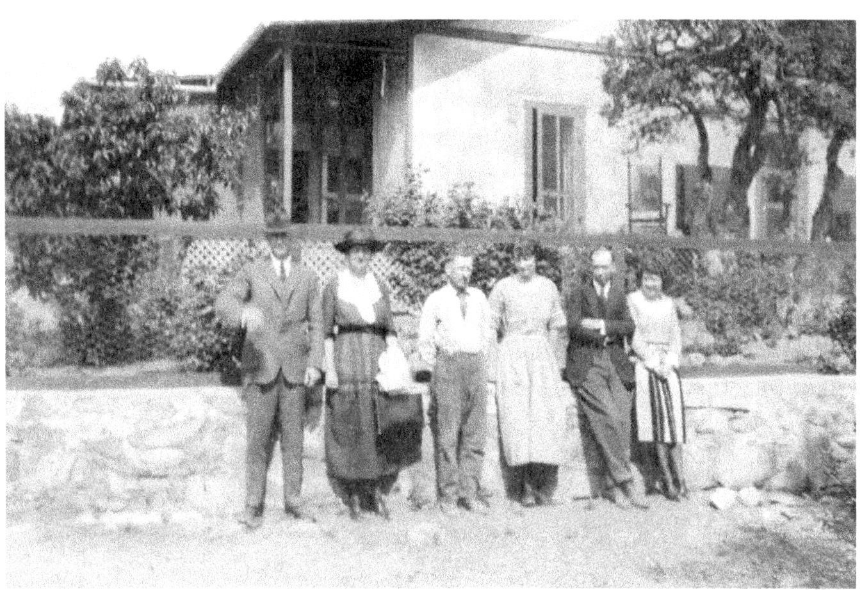

Triangle L Ranch visitors in its early years as a guest ranch. *Courtesy of the Triangle L Ranch Bed and Breakfast.*

had Mrs. Meta J. Tutt manage the ranch, and there were allegations that he was involved with her daughter. He later married a Scottish woman named Katherine. Trowbridge would eventually purchase the historic Triangle T Ranch in Dragoon for the elder Mrs. Tuff.

Trowbridge's letters to his mistress Margaret are still held at the ranch. Her responses to his letters, though, had been burned—at his request. "It's a mystery how his letters showed up at the ranch still intact. In reading the letters, it is clear that she blackmailed him, he was giving her money, buying her many things at Macy's, but at some point it all turns around, the tone changes." Trowbridge writes about his lost faith in humanity. "It appears that he had to pay her off for some mysterious reason," says Sharon. Ironically, Trowbridge's hypochondriac wife outlived him.

In 1978, Tom and Margot Beeston took on the restoration of the ranch buildings and grounds. The Beestons are credited with really getting the tradition of the bed-and-breakfast going again, observes Sharon. "They lived at the ranch for over twenty years until I came along."

Architecturally, the ranch today is well preserved and extremely beautiful. Whitewashed adobe buildings with red tin roofs and cobalt blue trimming dwell in the shade of giant oaks. Stacked-wood corral fences wind all around

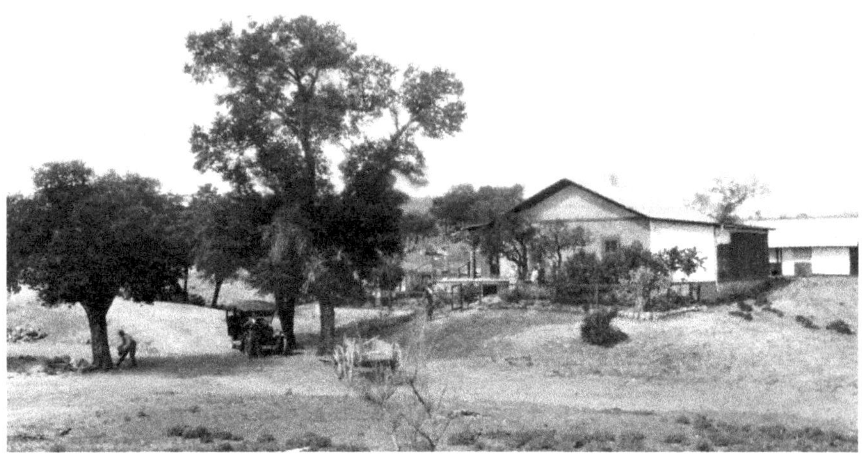

Oracle's Triangle L Ranch in the 1930s. *Courtesy of the Triangle L Ranch Bed and Breakfast.*

the property. "The old windmill still stands as a landmark from another era," comments Sharon.

The landscape of the Triangle L's property is equally as beautiful and in its natural state—desert flowers and cacti are everywhere. Habitat areas for wildlife, such as hawks, ravens, rabbits, roadrunners, quail, chipmunks and many species of songbirds, can be enjoyed during the day. At night, owls, javelina, bobcats and coyotes are frequent visitors with the occasional hint of a mountain lion among the boulders of the ranch's sculpture garden. In addition to the native fauna, Sharon tends a lovely iris garden, as well as a large community vegetable garden, and "we hope to provide some produce for the market and guests."

The ranch's most unique feature, though, is its outdoor art installations along winding pathways that stretch out into the high-desert ecosystem. It is unlike anything I've ever seen. The installations feature soundscapes that can be triggered by motion created by Michael Carroll. Because of overgrazing, much of the property was overrun with desert broom when Sharon first arrived. Sharon started weeding as though she was uncovering "buried treasures" of wonderful rocks and a beautiful environment. So she started a pathway that eventually expanded into an outdoor gallery and sculpture path where various artists have contributed installations or site-specific pieces.

"My challenge was figuring out how to balance historical integrity and nature while creating areas that can be utilized for community events, small

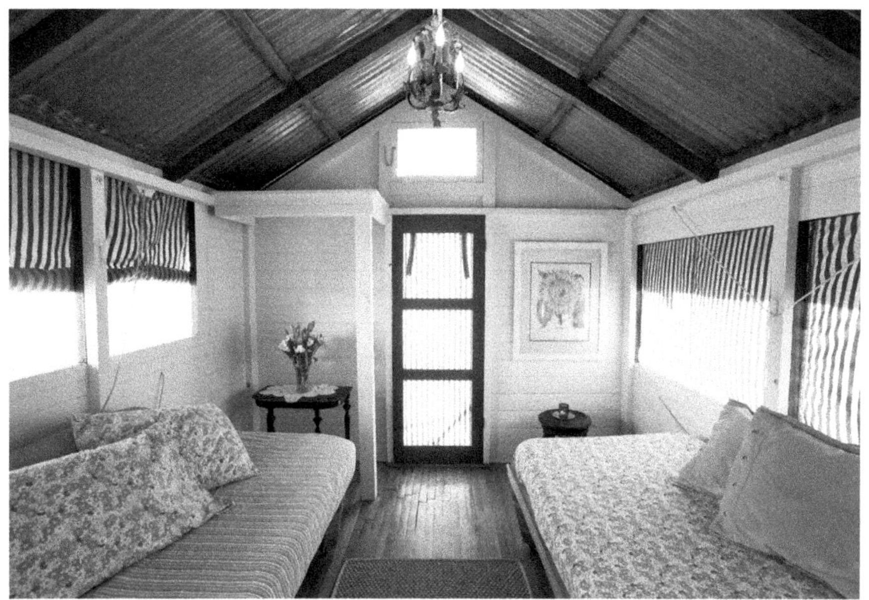

Guest accommodations at the Triangle L consist of private cottages carefully restored and charmingly furnished with antiques from each historical period of the ranch. *Photo by Jeff Smith 2010.*

gatherings, personal reflection and meditations," explains Sharon. "Guests and the community are welcome to go out into the sculpture garden, stroll and enjoy as a kind of park as well. We are constantly adding new environments for people to enjoy." Visit chickens, goats and dogs as an added bonus. Individual stays and group events such as weddings, retreats, reunions or art workshops of all kinds are now offered at the ranch.

"With so few public examples of Oracle's history," the ranch has become a real artistic hub for the community, with its cultural and entertainment events for the public throughout the year. GLOW, the ranch's biggest event with over two hundred artists and close to one hundred volunteers in attendance, takes place close to the full moon in September and features illuminated sculpture, multi-media art, music, performances and, of course, local food.

"My inspiration in running the bed-and-breakfast is this property, its buildings, history and the land. My mission is to blend art and history in this beautiful natural setting and have it continue to be a place guests come and enjoy. The guest ranch is just part of the package."

Contact Information: 2805 N. Triangle L Ranch Rd. Oracle, AZ 85623 Tel: (520) 623-6732 Tucson, (520) 896-2123 Oracle, E-mail: info@trianglelranch.com, Website: www.triangleLranch.com

Accommodations: Four private cottages, carefully restored and furnished with comfortable antiques, are available for individual guest stays and retreats. Much of the furnishings Sharon inherited with the ranch originated from each historical era of the ranch's history.

Chapter 11

White Stallion Ranch

Tucson

"Unwind in the cool clear evenings with stars that never seemed so brilliant, the silence broken only by ranch and desert sounds. A giant saguaro points its finger to the sky and the rugged mountains rise majestically in the background. The desert is quiet save for the gentle hoof beats as your horse carries you along the trail."

The Accidental Conservationists

These words from a White Stallion brochure in the 1970s still ring true today.

One of the first things that impresses is White Stallion Ranch's proximity to urban Tucson, and at the same time, its removal from it. Cowboys, cattle, hay, mud, cacti and a desert lush with greenery after a recent rainstorm await as you approach the ranch on a long dirt road.

This adobe cattle ranch was originally built in the early 1900s. A small part of that exposed adobe can be seen in the original ranch building's dining room as a testament to the century of history here.

Chicago liquor store owner Max Zimmerman bought the ranch in 1940. Zimmerman had moved West to become part of the once-vibrant guest ranching industry in Tucson. Rechristened MZ Bar Ranch, most likely after his own name, Zimmerman's tenure as a guest ranch baron was short-lived. The retired governess to the Dupont family bought the ranch in 1948. Over the next decade, she continued its guest ranch tradition and for a time even provided housing for Air Force personnel and their families.

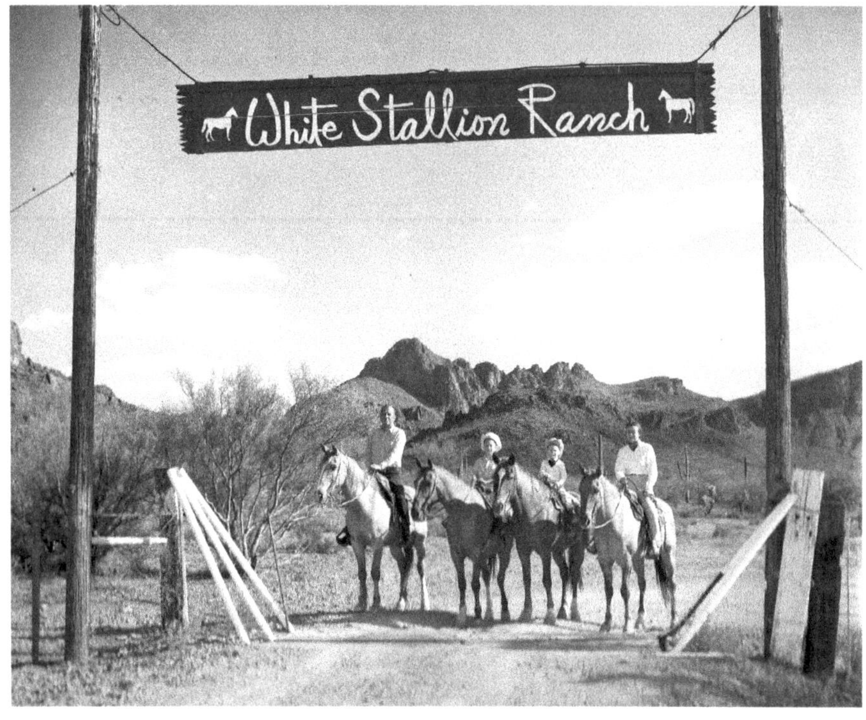

Located only twelve miles from downtown Tucson, White Stallion ranch guests enjoy trail riding through the foothills of the Tucson Mountains and Saguaro National Park. *Courtesy of White Stallion Ranch.*

In 1958, the ranch was sold to Cape Codders Brew and Marge Towne. Previous guest ranch vacationers in the area, they developed a love of the West and decided to try their luck as owners of the business. The Townes' favorite horse story was *The Black Stallion*, and they wanted to name the ranch after the novel. To their chagrin, they soon realized that the initials (BS) would be inappropriate, and thus, the name White Stallion was chosen instead.

By 1965, the Townes had sold the ranch to Cynthia and Allen True from Denver, Colorado. In search of warmer weather, Allen flew to Tucson, and "that was it," remembers his son Russell. By this time, the number of guest ranches in the area had shrunk from more than one hundred to approximately thirty as a result of the urban growth of Tucson. To avoid a similar fate and ensure an authentic experience for its guests, the Trues immediately began purchasing adjoining land around White Stallion, and the ranch grew to its current size of three thousand acres. To this day, the ranch is owned and operated by their sons, the second generation Trues, Russell and Michael.

White Stallion Ranch is now one of only two remaining guest ranches in the area. "The ranch's history is much like many dude ranches," says Russell True, owner and manager, who looks every bit like a rancher with his jeans and cowboy hat, with a firm grip and a friendly, plain-spoken manner. True grew up experiencing all dude ranching had to offer and knew that was what he wanted to do.

"People find the ranching lifestyle interesting but still have a lot of stereotypes of what a rancher is and what a guest ranch is in particular—they envision hokey, bad food, guests sleeping in barns. We have a lot of the same amenities and facilities of a resort, and yet we're not a resort," he says.

"Fresh air, lots of sunshine, activities if you want them, quiet times if you need them." White Stallion certainly looks like the prototypical Western guest ranch. "What likely makes us unique is how Tucson has grown close to us over the years." True's former school district and the usual city conveniences are nearby (a Safeway and cookie-cutter suburbs are a stone's throw away),

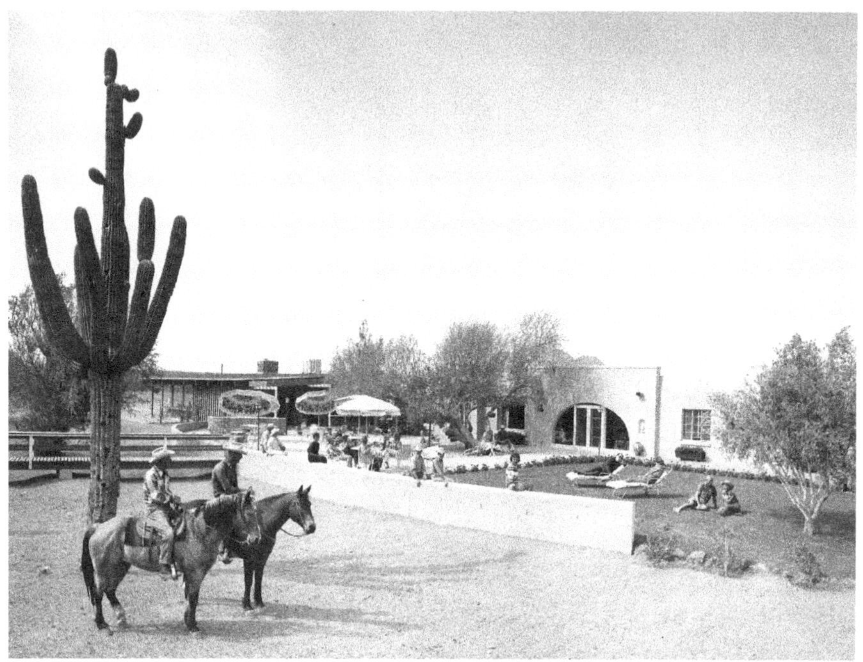

A working cattle ranch, as well as a guest ranch, White Stallion Ranch "religiously protects and preserves" its three thousand acres of land. *Courtesy of White Stallion Ranch.*

and yet they are separated by their mountains and truly jaw-dropping desert wilderness. "The best of both worlds, really," says True.

White Stallion has one of the largest herds of horses in Arizona of all kinds—quarter horses, thoroughbreds, Arabians, appaloosas, paints and mustangs of more than 120 and a large herd of approximately 150 cattle.

The daily operation of the ranch has changed "dramatically and not at all" since its early days, says one guest, comparing her first visit in 1965 to present-day ranch happenings. "It is still about horses, hats, hospitality and history," agrees True. The ranch's Longhorn cattle still come in from the desert for water every day with lots of feeding, branding, herding and check-ups to see about.

"The part that has changed is everything else. Our culture has changed and with it our ranch. People want to be entertained, educated and catered to. Standards have risen and tolerance for less has faded, if you want to stay in the business you adapt," says True. "We offer more options, more comfort, more entertainment and more educational programs, from astronomy to line dancing and everything in between." Despite the prominence of social media, thirty years ago, we as a society were more social, in True's opinion. In those days, a ranch vacation experience was more spontaneous—guests played cards or sat around the campfire talking into the night. "People made their own entertainment. They didn't want ranch staff around to interfere with that." Today's guests are more lesson-oriented and desire a learning experience on their vacations.

Horseback riding is still one of the keystone activities guests come to enjoy and can be done up to four times daily out in the open desert, along with hiking in the mountains, hayrides, stargazing, cookouts, team cattle penning and weekend rodeos.

White Stallion is moving toward a more energy-efficient future to become more self-reliant in the long-term. Solar panels produce "about 35 percent of our total usage" of power. "Our goal is to create an online monitoring system that guests can view. We replaced eight of the conventional water heaters with state of the art 'on demand' systems to further reduce our energy consumption," explains True. Additionally, air conditioners were all upgraded to "provide 40 percent more efficient service to our guests." Guests appreciate and support these endeavors.

The ranch also supports, preserves and "religiously protects" its three thousand acres of land "by virtue of our existence as a dude ranch," says True. Indeed, much of their three thousand acres would have most likely have been turned into another strip mall so frequently seen in Arizona. When

Hollywood at White Stallion

Because of its proximity to Los Angeles, Hollywood discovered White Stallion in its early cattle ranching days. Since the 1930s, the ranch's "movie pass" with its 360-degree views of classic Arizona landscape has appeared in numerous movies and on television, in commercials, rock videos, reality shows and publications such as *National Geographic*, *Playboy* and the *Sports Illustrated Swim Suit Edition*. A small sample of movies that were filmed at the ranch include the epic blockbuster *Arizona* with William Holden 1939, *The Last Round Up* with Gene Autry 1948, *The Last Outpost* with Ronald Reagan 1951, *A Kiss Before Dying* with Robert Wagner 1955, *Confessions Of A Dangerous Mind* with George Clooney & Sam Rockwell 2002, *Geronimo*, the T.V. Movie, and *How The West Was Won*, the T.V. mini-series and the western television show *High Chaparral*.

True's parents expanded the ranch in the 1960s, they inadvertently saved a significant amount of wilderness, becoming accidental conservationists. "We need the land to remain as untouched as possible, as unspoiled as it can remain."

True predicts that the ranch will change in the future, but those changes will be determined by guests. "We are where we want to be, where we envisioned it. Ten years ago, I never would have imagined we'd even have internet access on the ranch."

If history is any indication, the ranch will change in unexpected ways but stay True to its roots.

Contact Info: 9251 W. Twin Peaks Rd. Tucson, AZ 85743, Tel: (520) 297-0252, E-mail: rtrue@whitestallion.com, Website: www.wsranch.com

Accommodations: White Stallion Ranch's well-preserved forty-one rooms and suites, plus a four-bedroom hacienda or "main house," accommodate guests. A fitness center and sauna, fantastic redwood hot tub, tennis court and a large recreation room adjoining a movie theater rejuvenate and entertain.

Chapter 12

Elkhorn Ranch

Tucson

In these Purple Mountains Hospitality is Second Nature

A little over an hour's drive southwest of Tucson is a guest ranch where the journey there is a part of the fun and adventure of visiting this part of the state. You have to have a reason to travel to this remote part of Southern Arizona. The sleepy town of Three Points is the last outpost of "civilization," and then you encounter vast expanses of Baboquivari Mountains (sacred to the Tohono O'odham Nation) and pristine Altar Valley desert scenery as you take a dirt road up, up, up to the Elkhorn Ranch. Arriving here feels, in a way, like you have arrived in heaven.

Elkhorn Ranch is located at an altitude of 3,700 feet at the mouth of Sabino Canyon. At this elevation, the climate is soft, dry, warm and caressing. There are children playing and swimming, guests reading on the porch and bird life twittering. Informal and relaxed, the ranch and its environs also have a unique spirituality. I immediately feel lighter and more at peace. Stress evaporates from my shoulders, and I quickly forget my weekly "to-do" list.

The Elkhorn Ranch encompasses over ten thousand acres. Rich in historic significance, Elkhorn was originally part of the even vaster Otero cattle ranch. Reputedly in the cattle business longer than any other family in Arizona, the first Oteros arrived in the Arizona Territory from Spain in the 1700s. "Cattle king" Sabino Otero owned several ranches in the Tucson area in the 1800s, selling his brand to lucrative Eastern markets.

Sabino was known for his charitable works for many organizations, most notably helping to found Saint Mary's Hospital in Tucson. In addition to his philanthropic contributions, Sabino left his mark on the area in other ways. Elkhorn Ranch's Sabino Canyon (a more off-the-beaten-path, far less touristy alternative to Tucson's) is said to be named after Otero.

In 1945, Grace and Ernest Miller from Montana acquired the Elkhorn. Grace and Ernest got started with guest ranching in 1922 in Montana and were founding members of the Dude Ranchers' Association. While Grace focused on turning Elkhorn into a first-rate dude ranch, Ernest traveled throughout North America, from Alaska to Mexico, as a renowned hunting guide. Hunting trips familiarized the Millers with Southern Arizona. Grace and Ernest had been searching for many years to find an opportunity to operate their guest ranching operation year-round, she explains, "to take what they did up north and bring it south."

It was the Millers' colleague and friend Jim Converse, then owner of Tanque Verde Ranch in Tucson, who discovered a site for a Southern Arizona Elkhorn. At the time, the ranch was known as the Fresnal Ranch School, a boys boarding school that had closed in the 1940s. "It was an amazing opportunity because the infrastructure for a guest ranch was already there," says Mary Miller, the ranch's soft-spoken co-owner. "Most buildings on the ranch today date back to the boys' school."

Elkhorn has been continuously owned and operated by Grace and Ernest's sons and their wives ever since, now in the third generation. Few guest ranches in Arizona have remained family owned and in business for so long with such success.

What brings generations of guests back year after year to the ranch for their vacations? "The ranch is part of people's definition of home," says Mary, a "low-key, outdoor, peaceful, family-oriented place." The main buildings of the ranch are made of white stucco. Reading and conversation trumps TV out here. An extensive library with a large selection of titles on Southwestern history is the way to spend a lazy afternoon up here.

Guests get to know each other year after year and form lasting friendships. Sarah King, the Elkhorn's trail-riding guide who now lives at the neighboring ranch Kings Anvil (1894), started coming to the Elkhorn as a child. "The Millers do a fantastic job of welcoming people into their home," she says. "It's not cookie-cutter. The four tables in the dining room are the heart of the Elkhorn, very home-style and welcoming."

Horseback riding is the main activity that guests come to enjoy. The Elkhorn's 120 horses, including mares, colts and saddle horses, are raised

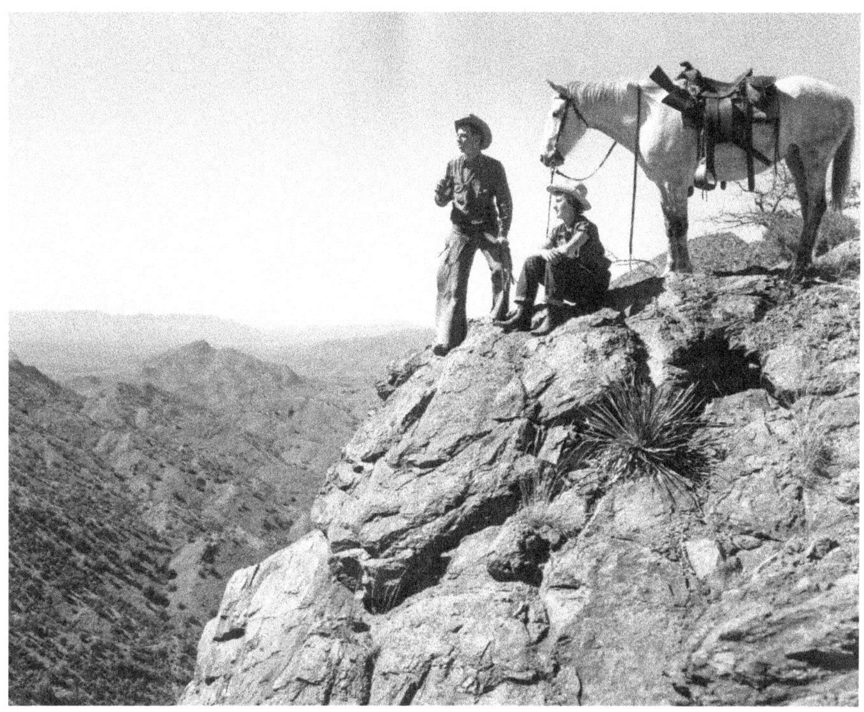

Taken in the 1950s, Elkhorn Ranch guests are "on top of the world" in the surrounding Altar Valley. *Courtesy of the Elkhorn Ranch.*

and trained at the ranch. Trail riding is offered during the Elkhorn's winter season of November through April (the ranch is closed during the summer months).

People of all ages have learned to ride or honed their riding skills on the ranch's many desert trails. It is quite a memorable scene for any "city slicker" to see all those horses grazing in the shadow of the mountains, their locational bells chiming.

Complementary to its world-class riding opportunities, hiking, tennis, a fifty-foot heated pool, bird watching, wild-flower viewing and picnics in the desert are a small sample of pastimes to enjoy at the ranch.

In addition to providing unique travel experiences, the Elkhorn is a working ranch that raises horses. Elkhorn's horses are uniquely suited to the rocky and scenic terrain of the Baboquivari Mountains, as many were raised on the ranch and all live on the ranch year-round. They are turned out nightly and all summer to graze in pastures surrounding ranch headquarters.

With state-of-the-art grazing practices to manage their land, the ranch cooperates with the Altar Valley Conservation Alliance, a group the Millers and their ranching neighbors started in 1995 with the vision to conserve the land for future generations.

Thanks to these ranches in the Altar Valley, the land remains a productive area for wildlife and for visitors to enjoy and experience. The unchanged landscape and ecology must have looked and functioned in very much the same way during the Otero family's ranching days.

One ranch guest puts it more simply—"I just really don't want to leave."

I have the same feeling as I begin my return to Tucson.

Contact Information: 27000 West Elkhorn Ranch Road, Tucson, AZ 85836, Tel: (520) 822-1040, E-mail: office@elkhornranch.com Website: www.elkhornranch.com

Accommodations: Guest rates are all-inclusive with stays of at least a week recommended. Guests stay in historic cabins spread throughout the property shaded by mesquite and unique Sonoran desert vegetation. Over the years, Elkhorn guests have lovingly contributed art and poetry to many of these accommodations. Home-cooked, nutritious meals are served buffet-style, and healthy, hot breakfasts of your choice can also be ordered.

Chapter 13

Triangle T Historic Ranch

Dragoon

Host to Everyone from Presidents to Prisoners

With an absence of pretension, location is big here. The Triangle T Historic Ranch is located in Texas Canyon, just off I-10 on 160 acres of oak tree–peppered land in the foothills of the Dragoon Mountains.

At 4,300 feet, the ranch's unique grassland ecology resembles Southern Africa. I almost expect to see a lion or leopard nestled in the tall yellow grass. A coyote jumps atop a large boulder and observes me with half-hearted interest. The ranch's wrangler and craftsman in a cowboy hat stands by the ranch's public restaurant and saloon. The smell of barbecued meat floats through the air.

Formerly called "Three Oaks" because of the (now-gone) high oak trees around the main ranch house, its earlier owners, the Seabring family, operated a fruit ranch. More than two hundred plum, peach, cherry and apple trees still remain. Built around 1922, the ranch became the home of Mrs. Meta J. Tutt, formerly of Oracle, and her daughter Katherine Tutt in 1928. It was Miss Tutt who designed the various ranch casitas and planned the ranch's artistic furnishings of precious stones for fireplaces, badger skins, Native American rugs and baskets. From just one red-roofed ranch house, the Triangle T gradually became a guest ranch of many buildings around this time.

The Triangle T's history is a kind of mini-timeline that mirrors events in Arizona and U.S. history. Texas Canyon was sacred ground for the HoHokam

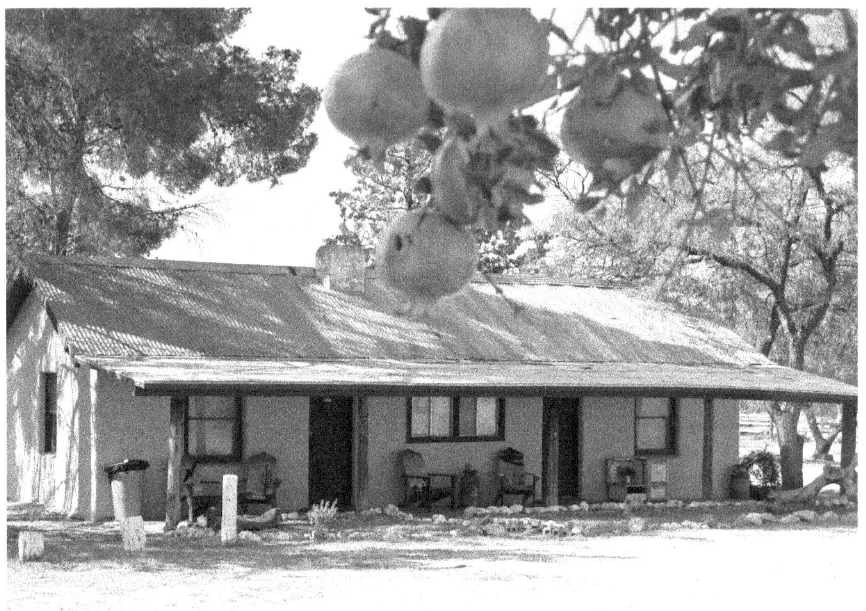

Texas Canyon's Triangle T Historic Ranch is a microcosm of Southwestern history and a mecca for world leaders in politics, business and the arts. *Courtesy of Sherry Harrison.*

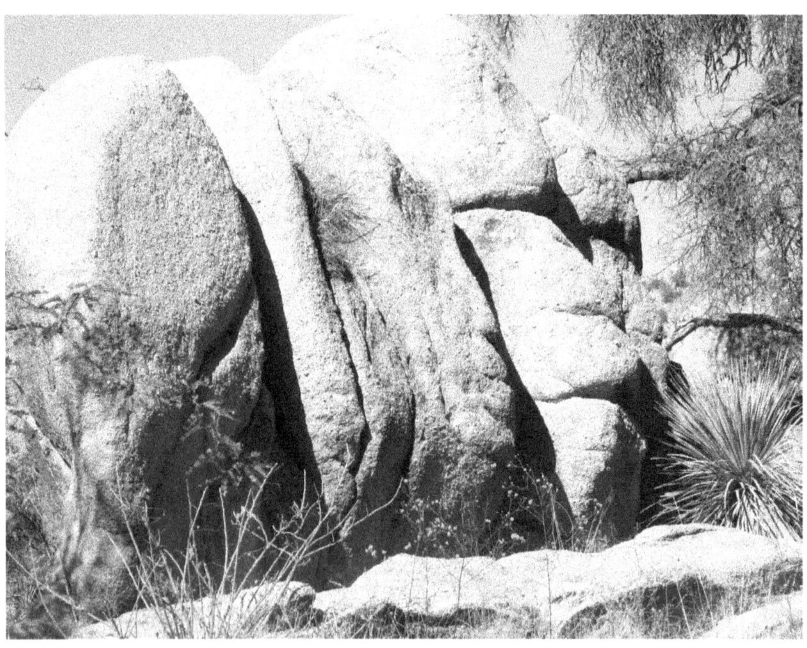

The Triangle T's large amount of massive, scattered boulder formations are the result of a prehistoric volcanic eruption. *Courtesy of Sherry Harrison.*

Indians prior to the 1800s, and later, the actual ranch site became Chief Cochise's winter camp. In 1874, Cochise was buried in a secret ceremony somewhere high in the Dragoon Mountains.

During the 1930s, Clark Gable, Gregory Peck and the Lone Ranger vacationed at the ranch, and in 1934, John Wayne filmed *West of the Divide* here. A 1930 ranch brochure lists the home addresses of the Vanderbilts, the Rockefellers and other prominent families who were friends of the Tutts and guests at the ranch. In those days, guest ranches provided such references to prospective guests. "There was and is a special energy about the place," says the ranch's current owner, Linda Kelly. "In its day, this was the place to be."

After the bombing of Pearl Harbor in 1942, President Roosevelt secretly transported the infamous Consul General Nagoa Kita and his staff from Hawaii to the Triangle T to be held as war criminals. Decades later, the ranch's maid found a letter written in Japanese but accidentally threw it away. Who knows what intriguing communique was lost forever?

During the early 1960s, it is rumored that army general "Black Jack" Pershing, who invaded Mexico to capture Pancho Villa, was a ranch guest. During this time period, President John Kennedy; Dr. Silver, advisor to NASA; Steve McQueen; Walt Disney; and many authors, artists and historians flocked to the ranch in a kind of renaissance movement. Some of the older Disney cartoons resemble the scenery at Texas Canyon. Paul Scott Mowrer, New Hampshire's first poet laureate, wrote "Texas Canyon, Arizona—A Writer Fantasy" while staying at the ranch.

The original *3:10 to Yuma* was filmed at the ranch. Preserved for guests and the public to see today are the film's ranch house where Glenn Ford was held prisoner. The immediate area of the ranch has served as the location for many other films and television productions over the years, including the TV series *Young Guns* and movies *Geronimo* and *Tombstone*. An A&E special *Psychic Kids*, *Man vs. Wild* and several TV shows, commercials and music projects were filmed at the ranch in 2008 alone.

To this day, movie and television production planning goes down in a small dining room adjacent to the main dining hall. "The film group from the UK that came to the ranch last October stayed for 2 weeks and had the time of their lives. At the wrap-up party, we brought in a Western band, and they drank and sang all evening. It was quite a cast party," remembers Linda.

Like many historic properties in Arizona, the Triangle T has its fair share of ghost lore. The sun is always shining, but there's a dark, eerie undercurrent here. Dead Indian remains almost call out from the ground. After all, it served as an internment camp for Japanese and

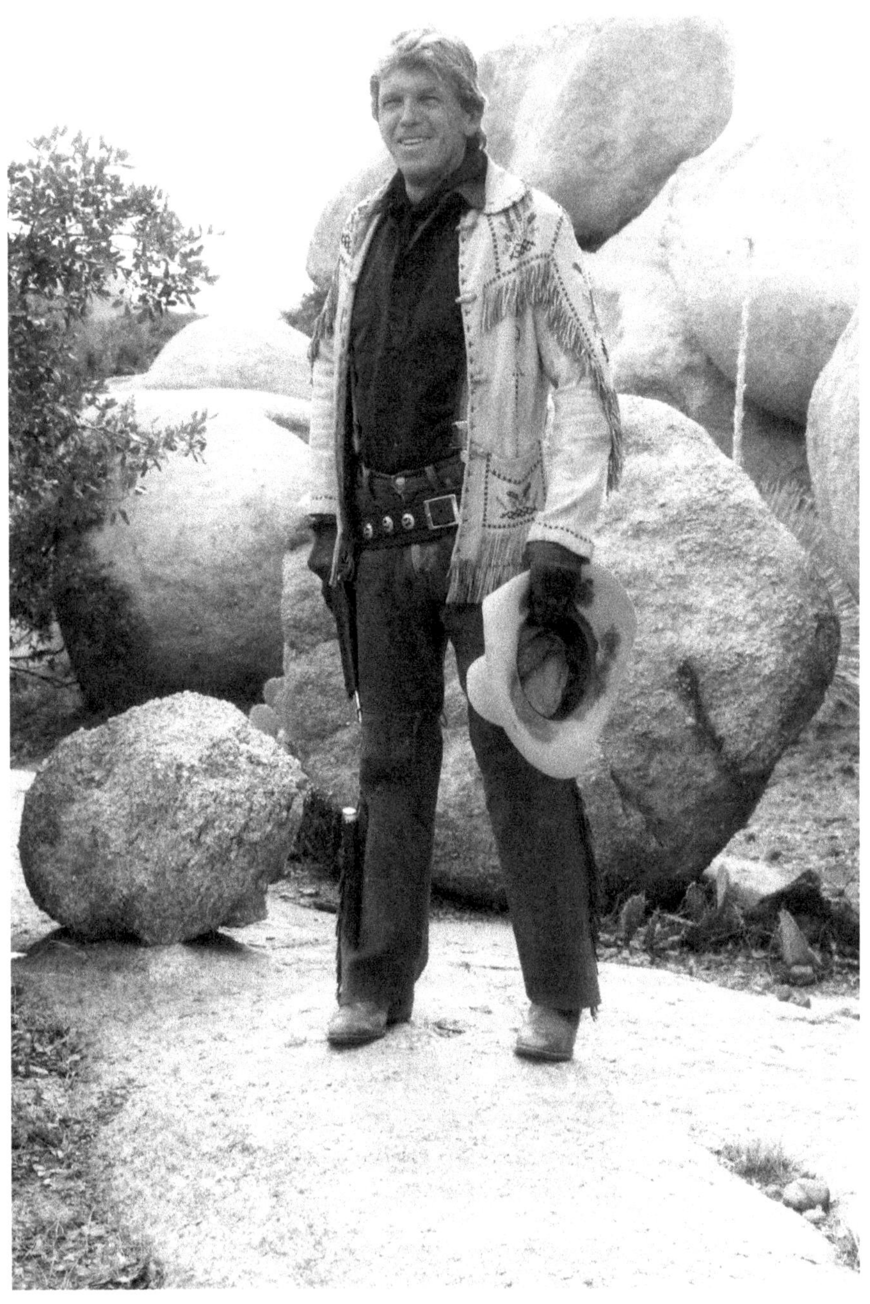

Publicity shot of Triangle T ranch guest Jody McQueen. *Courtesy of Sherry Harrison.*

Italian prisoners who were sent here because of the isolation. During the 1940s, they were trapped on the ranch with nothing around for miles but a harsh, unfamiliar environment to travel through should they escape. A tennis court was built just so that these prisoners could exercise and can still be seen at the ranch. "Someone was banging on the door last night around 1:00 a.m., [but] when we checked, no one was there," claims one British guest.

Despite a restless ghost or two, guests come from Europe, Canada and all over the United States to experience this unique Southwestern getaway. "Most of our guests learn about the ranch from word-of-mouth. They are all looking for a quiet, serene, relaxing and re-energizing time away from their hectic lives" says Sherry Harrison, a kind of ranch historian. "What's special about the ranch is that it's hosted everyone from presidents to prisoners over the years." It has indeed been a disparate group of guests.

The Triangle T's popular all-inclusive three-day to weeklong riding programs that include advanced rides, off-the-ranch rides, training and arena competitions enable them to do this. The Amerind Foundation archaeological research center and museum is next door. This high-

Blueberry Pie

from the Triangle T Historic Ranch

Filling:
12 cups blueberries
1¼ cup sugar
⅓ cup cornstarch
½ cup lemon juice
Mix together and pour into your pie shell.

Crumb Topping:
3 cups of brown sugar
5 cups of flour
1 tablespoon cinnamon
1 stick of butter
Mix together and use as the topping for the three pies. Bake at 350 degrees for one hour.

desert environment and mountainous terrain offers myriad opportunities for hiking, bird watching and day trips to historic Bisbee or Tombstone. Catch one of their annual events such as July's Garlic Festival, the Desert Archer's "Rumble in the Rocks" in the spring or Christmas Gala.

"All our guests are attracted by the spectacular scenery and authentic old West ambiance," says Sherry. The ranch resembles a big extended family (they recently held a funeral for a former guest) of world-famous and mere mortal guests. Jody McQueen, Steve McQueen's brother and stunt double, recounts one of his fondest memories of the ranch: "I and a couple of stunt-actor friends started to act out a fight scene in the unsuspecting busy Rock Saloon at the Triangle T. One of the actors accidentally went through the picture window as I said: 'It wasn't candy glass.' Luckily nobody was hurt… but we all felt it was worth the looks on the faces of those who were not in the know."

Soon after, Jody's career took him all over the country, and while he had not been back to the Triangle T for a while, he still remembers the hospitality of the locals. "The people of Southern Arizona are down to earth, genuine and will give you the shirts off their backs. Coming back to the Triangle T was like coming home."

McQueen is now one ranch guest here to stay.

Contact Information: P.O. Box 218, 4190 Dragoon Road, Dragoon AZ 85609, Tel: (520) 586-7533, E-mail: ttgr@earthlink.net, Website: www.triangletguestranch.com

Accommodations: Lodging options include the ranch's historic casitas or bunkhouse. Full or partial RV hook-ups are available as well as plenty of space for dry camping.

Chapter 14

Cherry Valley Ranch

Oracle

Along Oracle's Wildlife Corridor, the Dream of the West is Alive

More than anything, Cherry Valley Ranch Bed and Breakfast looks like someone's home. And it is and has been for the last fifty-six years or so.

Located in the foothills of the Santa Catalina Mountains with views of the San Pedro Valley, Cherry Valley Ranch Bed and Breakfast is positioned on oak woodland terrain and desert grassland with ancient granite boulder outcroppings. "When you pass the modern steel gate to the ranch, you are leaving behind the hectic pace of today. You are immediately struck by the beauty of this land. Wild flowers, in season, cactus of all sorts, neatly trimmed trees and rustic fences form days gone by," described recent ranch guest David Brown.

Cherry Valley was a popular picnicking area in Oracle's early days and, even today, is a traditional acorn and bear grass gathering site for the Tohono O'odham and Apache. The name of the ranch originates from the edible wild-cherry trees that grow on the property.

"Cherry Valley Ranch and the establishments of the bed-and-breakfast is a microcosm of the industry of hospitality within Arizona ranching," explains owner Susan Woodruff. "Mining, health-seekers and ranching sustained early Oracle, and soon people in Tucson sought to escape the desert's summer heat—first staying at the Mountain View hotel in the center of Oracle, then spreading to purchase adjacent land for homesteads."

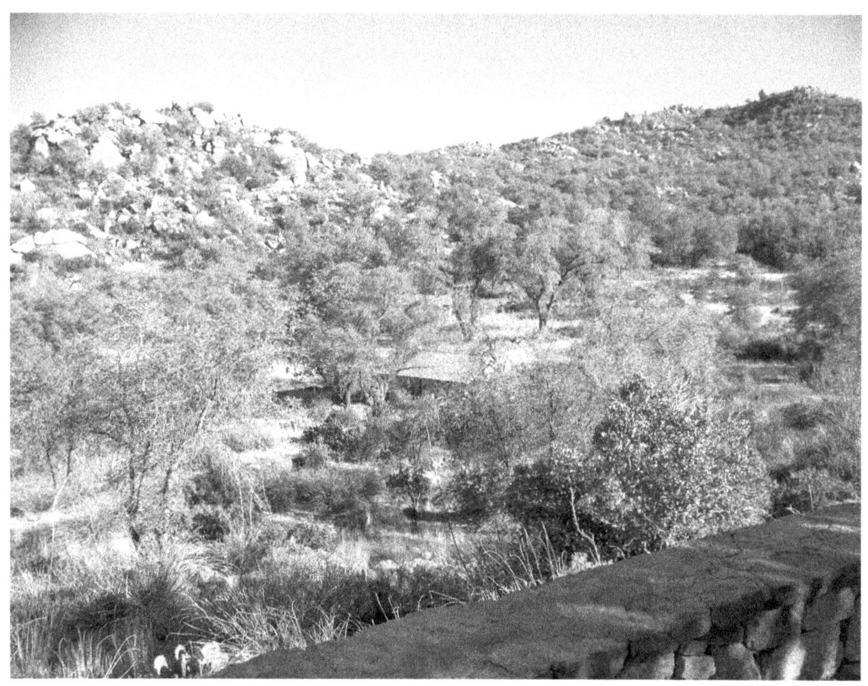

Located in the foothills of the Santa Catalina Mountains, Cherry Valley Ranch Bed and Breakfast's one hundred acres of oak woodland terrain is an essential wildlife corridor for mountain lions, black bears and scores of other native wildlife.

In the late 1930s, Colonel William Bacon and his wife acquired the one hundred acres that comprise Cherry Valley Ranch as their home. A former army officer prior to and during World War II in Tucson, Bacon also secured a grazing allotment on neighboring Coronado Forest Service land after he purchased the ranch and left his "legacy of silver trimmed saddles, bridles and personal chaps," observes Patricia Woodruff, Susan's sister. "Their finely-honed good taste from the east, and their love and appreciation of the arts of the west, melded Cherry Valley Ranch into a beautiful home."

Designed by prominent Tucson architect John "Ginger" Smith, Cherry Valley's ranch house was constructed by a family of stone masons from Mexico using granite quarried over a two-year period. In the late 1940s, the ranch was sold to Henry Sanford of Connecticut. Having vacationed several times at the Circle Z Ranch in Patagonia, the Sanfords "had always fancied their own ranch where Mr. Sanford could pursue a writing career," says Patricia.

When, in 1956, Frank and Mary Woodruff and their seven children were transferred from Kennecott Copper Corporation in Nevada to Ray Mines Division in Hayden, the ranch's present ownership was born. Oracle's Cherry Valley Ranch was the closest home available at that time with room for such a large family. Oracle was even more rural then, and Susan remembers being sent to the store in town on horseback in those days.

Frank Woodruff was "a bit of a romantic," according to his daughters and, shortly after the move, planted the seeds of opening a guest ranch in the tradition of earlier guest ranches of the area. However, the "overwhelming task of raising such a large family snatched the dream from completion" until 1994 when his children opened a bed-and-breakfast at the ranch.

The ranch's most significant feature is that it is located on an important wildlife corridor. Due to preservation by Arizona State Parks and Cherry Valley Ranch Bed and Breakfast, Cherry Valley Wash allows easy movement of wildlife between the San Pedro River, the Galiuro Mountains and Mount Lemmon. A mountain lion was recently captured and radio collared right on the property. "Cherry Valley Ranch has become, by its lack of development, a mainstay of Oracle's original habitat," says Patricia.

Guests with a love for the outdoors can enjoy many activities during their stay—hike the Santa Catalina Mountains' many trails or try "eco-trekking" with local llamas. Parties of twelve or more can enjoy horseback riding at Oracle's Triangle Y camp. "There's always an adventure here," says Susan. Today, it's the rodeo horses here for a rest that bring the adventure as they block the dirt road up to the ranch.

Ranch guest and avid hiker Bonnie Halchin-Smith, a teacher from Columbus, Ohio, found the blooming wildflowers, phenomenal night skies and wildlife viewing—a rattlesnake and Gila monster, as well as much bird life—most memorable. "In the morning, my husband, Susan and I sat on the porch over breakfast watching snow fall. Susan was sitting with a bird field guide, pointing out various birds and said 'I never take it for granted.' Her love and loyalty (to the ranch) can't help but make you feel more at home. The amenities that a hotel may offer can't possibly equal that," says Bonnie.

Guests come from all over the world and return year after year. Countries such as Japan, Switzerland, Germany, Ireland, England, Argentina, Mexico and Panama have all been represented over the years. "Many people come to visit their family in the area, others are attracted to the Oracle community and its art and some religious groups visit for a spiritual uplift with nature," says Susan. One thing unites these diverse groups. "Everyone comments on the beauty, peace and quiet," or, as one guest writes, the "lovely respite and magical coming home."

Contact Information: 2505 Mount Lemmon Road, Oracle, AZ 85623, Tel: (520) 896-9639, E-mail: cherryvalleybnb@yahoo.com, Website: www.cherryvalleyranch.com

Accommodations: Guests can choose to stay in two large, exquisitely furnished bedrooms in the main house of the ranch with a fireplace, original fresco type art, antique tinwork and reproduction Spanish Colonial furniture or two private cottages in the valley below. Savor oatmeal, fruit, tortillas, beans and rice and Susan's fresh pecan pancakes over breakfast.

Chapter 15

3C Ranch

Oracle

The Not So Little Ranch That Could

The 3C (Columbia Cattle Company) Ranch is a historic guest ranch situated between the Santa Catalina and the Galiuro Mountains in Oracle with a tumultuous past. When cattle ranchers first came to the area in the 1880s with its abundance of water, the carrying capacity of the 3C's land was over 610 cows. By 1940, this working cattle ranch had grown to encompass thirty-six thousand acres. It was miner Mary West who had unified the ranch from several smaller properties. The bulk of the acreage was once part of the Bill Huggett Ranch, with some land coming from smaller ranches in the area—the Peppersauce, the American Flag, White House and VY.

The ranch was owned by Theodore J. Vaitses, a Massachusetts legislator, since 1958 and eventually became a magnet for the Hollywood and political set, with Mae West, President Nixon and many others as frequent guests of the ranch.

In 1966, the ranch was sold to Mr. and Mrs. Delmar W. Rowe of Bloomington, Illinois. A hotel owner and investor, Rowe reportedly paid over $1 million for the property. "Red Woverton managed the ranch in the 1970s and also owned most of the wagons and stagecoaches at the Old Tucson Studios. John Wayne is even said to have played poker at the ranch from time to time," says Adrian Darimont, the 3C's current owner, of the ranch's historical tidbits.

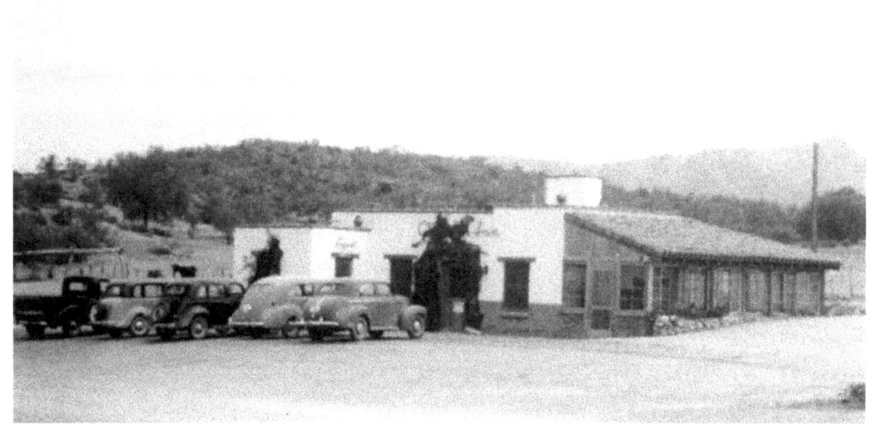

The Historic Oracle Inn Steakhouse and Lounge during the 1940s, now a gathering place for 3C Ranch guests. Here, the ranch's owner prepares and serves family recipes from the Black Forest area of southwest Germany. *Courtesy of the Oracle Historical Society.*

After the death of her husband, Mrs. Rowe inherited the ranch and ran it for a time. She even tried to turn the 3C into a guest ranch (without success though). In 1978, Mrs. Rowe deeded most of the ranch property to the Arizona Title Insurance and Trust Company. Then in 1982, the ranch was sold to the Tucson-based Niles Thim Corporation, a real estate investment firm.

Throughout much of the 1980s, the ranch was in danger of being lost to development. In 1984, Niles Thim proposed that the ranch be tuned into a two-thousand-home development and shopping area with the main ranch house converted into a resort. Oracle residents, fearing "Frankensteinian growth," opposed the plan, citing a lack of sufficient water to support the planned community. They felt the recreational value of the area, as well as the environmental quality of the land, would suffer as well. Another plan to turn the ranch into a big-game shooting preserve not long after met with similar resistance.

Then the 3C's luck would change and so would its destiny. Acres of the 3C Ranch were purchased piece by piece by the Darimont family during the next decade. "The 3C purchases actually started with my parents purchasing the first acreage and followed by several more purchases thereafter," says Adrian, a real estate broker. "I found out about the original parcel for sale in 1992 from a broker friend of mine and sold it to my father. He then kept buying parcels as I did research finding more adjacent to what he already had. I always told my clients, my dad being one of my best clients, 'never fall in love with your investments.' Well my dad did, so he kept buying and never sold any of it. As years went by, I purchased the last two parcels, because I also fell in love with the area," describes Adrian. "It was my parents that deserve the credit for making this ranch what it is."

In 2004, the Darimonts began restoring and rebuilding the dilapidated ranch. As a result of painstaking efforts, the 3C Ranch's three fully renovated guest houses and modern horse facilities are open for business year-round. The Darimonts also own the Historic Oracle Inn Steakhouse and Lounge right in town. "People that stay at the ranch always have a great place to eat and great Western entertainment on weekends," says Adrian. "It is also the only steakhouse that serves original German food on the second weekend of every month. That is the busiest weekend of the month, serving my family recipes from the area of the Black Forest in southwest Germany, where I am also from. I get to visit some of the customers that speak German and brush up on my language skills that are getting a little rusty."

Situated on over five hundred acres, shade trees and well-manicured lawns enclose the ranch houses. Its landscape is secluded, peaceful and expansive. The ranch grows fresh vegetables that are utilized in guest meals and for ranch use. Numerous trails wind through the property with the possibility to spot wild turkey, blacktailed and white-tailed deer and several varieties of quail. The ranch's old mines, long closed, have produced gold, silver, copper, lead and zinc for prospectors. Gold was once panned in its streams.

Guests can enjoy hiking, horseback riding during special events and mountain biking. Explore old silver mines, bird watch or get a massage after a long day's exertion. Weddings, retreats and customizable vacation packages are available for prospective guests. The ranch has finally found a permanent home and purpose in the strong guest ranching community of Oracle.

Contact Information: Tel: (520) 896-3533, E-mail: ADarimont@aol.com, Website: www.3cranchaz.com

Accommodations: The ranch's three guest houses for overnight lodging include a four-thousand-square foot "Ranch House" designed by a prominent Boston architect in the 1960s, as well as the "Bunk House" and the "Nixon House," all with charismatic Western decor. Other ranch amenities include a jacuzzi, plasma TV and more than a few scenic views of valleys and mountains. Guests can board their own horses in the ranch's stables and corrals.

Chapter 16

Crown C Ranch

Sonoita

Where Time Stands Still

The beautiful grassland habitat in Santa Cruz County, east of Tucson, is a sight that I never tire of. Oak trees romantically dot the soft, gentle hills of pale yellow native grass as I pass the area's ranches and farms.

Located halfway between Sonoita and Patagonia, the Crown C Ranch is so low-key, it's easy to miss. Arriving on ranch property, at first you notice the silence. The only sounds are bird and insect life and the breeze blowing through the trees. Then, a friendly border collie appears and breaks up my reverie.

The Crown C is among the numerous ranches in the area that were split off from the massive Empire Ranch. The present day Crown C was formed in the mid-1930s. Its previous owner, New Jersey–born Blake Carrington, named the ranch "Crown C," which also served as the ranch brand—a C with a crown on it. The adobe ranch house was built during Carrington's time.

The ranch has remained a working cattle ranch. Blake and Jane Carrington met and married when Jane was a summer guest at a neighboring guest ranch. Chicago-born Jane was heavily involved with the ranch community and belonged to the local Elgin club, as well as Cowbelles, an old organization of ranch women that supported ranch activities. "Life on the ranch could get lonely, and this group was a way for them to get together to support and understand each other's concerns on the ranch, the concerns of women's work of the day," says the Crown C's current manager and part owner, Sidney Franklin, also a friend of Jane's.

The Carringtons invited the Franklins over to their ranch to participate in ranch activities. "We used to help work cattle on the ranch at times. Whenever someone had some work to do on a ranch, neighbors would pitch in," she says, commenting on the interdependence of ranch life.

Blake Carrington died in 1963, and Jane ran the ranch afterward with the help of foreman Dick Jimenez and his wife, Eva. After Jane died in 1979, the Franklins took over the ranch in early 1980. "Jane was a friend of ours, and we had an interest in the work. My husband developed the ranch is ways that were sensitive to the natural environment," says Sidney. "Most ranchers are good stewards of the land because the green is in your ground…It's a combination of love of the land and practicality."

The Franklins are from Pennsylvania originally, and Sidney Franklin had previously worked on cattle ranches in New Mexico. Her interest in ranching really took hold during this time. Sidney loved the wide-open spaces and mountains. A lifelong horseback rider and fantastic conversationalist, Sidney learned the ins and outs of the ranching business by working in New Mexico. She says she "got some experience but not even the size of a pinhead of what you need to know to run a ranch."

"My jobs were always indoors, because women weren't hired to do cattle work in the 1950s. But that's what I really love to do. It is a great thrill." Sidney was an exception in her day and was included in all ranch work, indoors and out.

The Franklins first came to Arizona at the end of the 1960s. "This area of Sonoita and Patagonia was just ranch people, all the development hadn't started. It was quite different." Many homes had not been built yet, and there were far fewer people. "At first, in ranching, it was a smaller group of people with cohesive interests, and now there's a greater diversity of interests and politics," she says.

The Franklins have made very few changes to the Crown C. "Changes have gone on all around us, but we've stayed in the past. We haven't been doing so much modernization." People advise her to modernize, but she just shrugs them off. Born in the 1920s, Sidney wants to maintain that family "get together" feeling and keep the ranch buildings true to that era.

The original ranch house is more than sixty years old and made of handcrafted adobes made on-site. "Everyone loves it here, there are no pretenses, no fancy stuff," she says. What little has been added was kept in the spirit of things. A mix of Mexican, Spanish and European influences, the decor is elegant and understated. Pillowcases are branded with the Crown C brand. The living room's library selection provides guests with sightseeing ideas in the area. Their old-fashioned pantry adjacent to the main dining room is from a more gentile era of entertaining. I almost expect a butler to emerge serving tea to guests.

Jane Carrington is still a "part of what's in the house, her feel is still there, the comfortable ambiance is hers, that feeling of belonging. She was a great hostess," says Sidney. An old black-and-white photograph showing Jane in a kind of Western fashion show gives a face to the feel.

More of a retreat instead of a true guest ranch, guest activities and entertainment are of the guests' own making. The vibe is "come, relax and do your own thing." Birders "see more birds here than anywhere they have been," and deer, javelina and the occasional bobcat and mountain lion are other ranch visitors.

The ranch's many return vacationers all comment on this "little slice of heaven" and the "pursuit of peace" during their stay. "The appeal is the beautiful setting, lovely authentic house. We've kept it essentially as it was, rather than jazzing it up and the many old furniture things and effects from the previous family," agrees Sidney.

Guests can bring their own horses, and the ranch provides stables and sustenance. With no scheduled activities or pressures to be up at any particular time, the privacy and peace here is unparalleled. Enjoy sunrises over the Santa Rita Mountains, views of the San Rafael Valley (a short drive from the ranch) just before sunset, black nights and star-filled and smog-free skies. Other unique features on-site include hiking trails, the ruins of historic Fort Crittendon and swimming facilities. The shops and restaurants in nearby Patagonia and Sonoita are minutes away by car.

Cattle can be seen out and about. This is a working cattle ranch, after all, and guests are welcome to observe the goings-on. But on a typical day, there is really not much going on. The ranch raises Angus and Hereford beef cattle, grass-fed. They sell their calves every year for meat.

"People come here to be themselves. I am not lurking around all the time (the Franklins live on a different part of the property). When you rent the house, it's yours. It's like a home away from home."

Contact Information: PO Box 505, Sonoita, AZ 85637, Tel: (520) 455-5739, E-mail: info@crowncranch.com, Website: www.crowncranch.com

Accommodations: Open year-round, the Crown C Ranch has room for up to eighteen guests. Reserve all six bedrooms in the main house, one of the house's spacious wings or a two-bedroom guest house. In all accommodation options, guests can relax in front of a stone fireplace with superb views of the landscape and mountains.

Chapter 17

KENYON RANCH

Tumacacori

"We have exited the West of hardscrabble struggle for survival and entered the West of relaxation and rejuvenation amid awesome scenery and clear desert air."
—Alan Hess

THE EXTENDED FAMILY OF THE SANTA CRUZ VALLEY

Just south of Tucson, Kenyon Ranch sits on high, rocky land in the foothills of the Tumacacori Mountains. At 3,400 feet, this peaceful ranch overlooks the historic Santa Cruz Valley. Since Father Kino's time, cattle have been raised around Tubac. Into this background of ranching arrived Easterners who had a lasting impact on the community. William T. Allen from Cincinnati, Ohio, had spent a few winters vacationing at the nearby Circle Z Ranch in Patagonia to relieve his asthma symptoms. Finding the climate conducive to work, this experience inspired Allen to buy land in nearby Tumacacori and start a guest ranch. Allen and a few of his close friends were soon constructing a complex of stone buildings from the ground up to house future guests.

Allen's letter to his daughter Vera dated April 14, 1937, writes: "Well I've gone and done it—I've bought us a western home...you can call it a ranch for we do have one hundred acres and it is right in the middle of real ranch country with a view that can't be beat—extending in all directions."

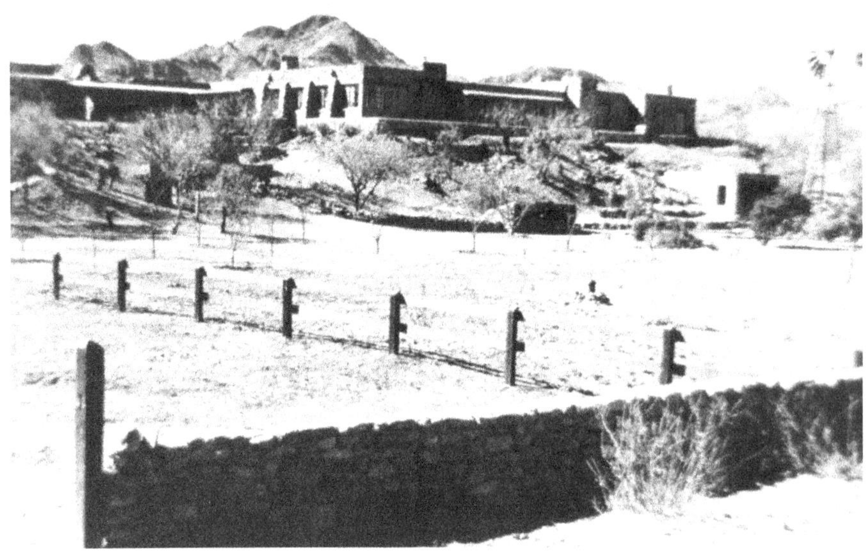

Opened in 1937, in its heyday, Kenyon Ranch was one of the most widely known western guest ranches in the country. *Courtesy of the Tubac Historical Society.*

Allen's wife, Marcella, or "Sis," soon joined him from back East. Soon, 100 acres grew to 1,700. "We loved it dearly and gave everything we had to that ranch," said Sis in an oral history interview years later.

Officially opened to guests that same year, the ranch was named Kenyon Ranch after Bill Allen's Ohio alma mater, Kenyon College. At first, there was only a four-room house on the property. The complex soon grew to include eleven separate cottages, a pool, a ramada, game rooms, stables and croquet and horseshoe courts. Horseback riding was its most popular guest activity. Trails led out from the ranch in every direction with no fences in those days.

Open for business from November to May, the ranch and its warm climate attracted prominent businessmen and their families as a popular winter vacation spot. From 1939 to 1974, Kenyon Ranch was one of the West's most widely known guest ranches. Although the ranch was featured in newspapers and magazines nationally and internationally, word of mouth was sufficient advertisement. Kenyon provided all guest services for only ten dollars per day in those early years. Food at Kenyon was always a major

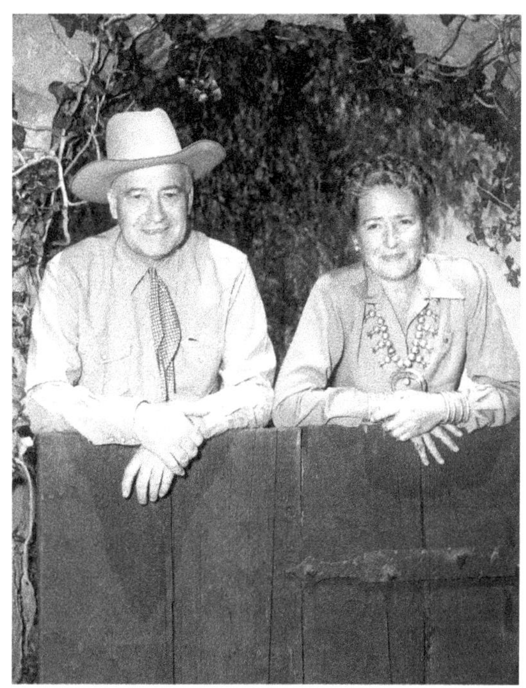

Left: Kenyon Ranch founders William T. Allen and his wife, Marcella, or "Sis," dedicated their lives to the ranch. *Courtesy of Vera Weintraub.*

Below: Kenyon Ranch guests relax by the pool in the 1950s. The ranch was like a "summer camp for grownups," says Bill and Sis Allen's granddaughter Lynne. *Courtesy of Kenyon Ranch.*

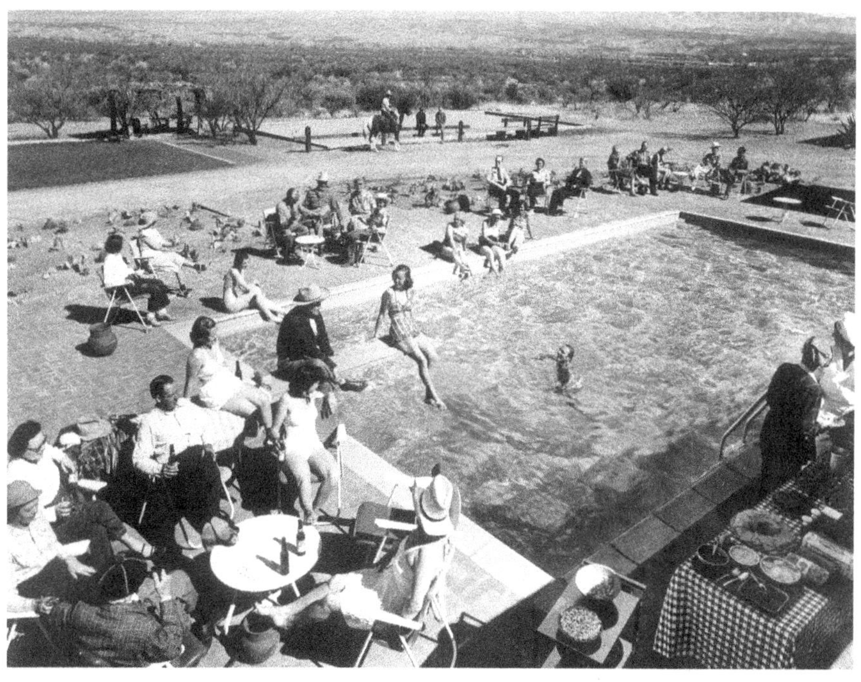

drawing card. A standing cow provided fresh butter and milk. Trail cookouts were popular—one infamous cookout even featured javelina, though many guests opted for chicken instead. The hunters ate the javelina.

With first-class meals, unsurpassed hospitality and Southwestern ambiance, guests soon became so regular, the same "core of people" year after year that many eventually purchased property in the area. Bill and Sis became founding members of the Tubac Center of the Arts, and many Kenyon guests contributed to building the organization.

Bill and Sis's granddaughter Lynne Thompson Rugg grew up at the ranch and remembers the day Cary Grant arrived at Kenyon for a stay. "I wondered why all the women were so a twitter all of the sudden?" Afterward, they were stunned that he was "so short." Other guests included Ricky Nelson, and Lucy Taliaferro and her husband, both professors at the University of Chicago. Lucy's brother's work on the construction of the first atomic reactor led to the construction of the first atomic bomb.

Over the next forty years, Bill and Sis maintained their guest ranch with panache. "It was a time and a place that could never be duplicated,

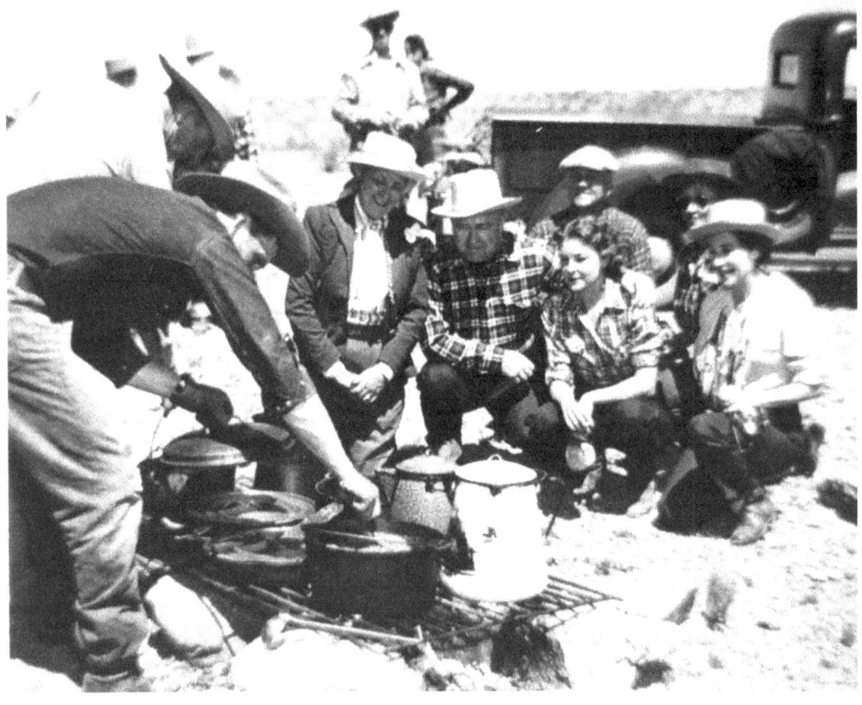

A family-style Kenyon Ranch breakfast ride in 1946. *Courtesy of the Tubac Historical Society.*

a fascinating time to be alive," says Lynne. In 1962, Bill Allen died, and Sis and her daughter-in-law Ann Nichols kept the ranch going for another twelve years. A Mr. Harry Pollack from Tubac bought the ranch, and later his widow, Penny, ran the ranch as a private residence.

Thereafter, each time that Kenyon was sold, Vera Weintraub feared that developers would buy the land and tear down the old buildings. "All that land belonged to my grandfather, Bill Allen, called 'granddaddy dear' by me, and I couldn't bear the idea that it would be torn down for housing," she says.

Vera visited the ranch in 1955 when she was twelve, seeing it for the last time when the ranch was running at full steam. Vera remembers BBQ lunches at the pool and riding with the cowboys cantering across open land. "At the time, there were few houses in sight and no highway. We used to go down to meet people on our horses at the gate," she says. "When we were very young, my mother would bring our schoolbooks, and we would spend the month of November living in the casita directly across from the old bell. These were some of the happiest days of my childhood. Our last two visits as a family were over the Christmas holidays, which were very special at Kenyon. Sis went over the border to shop in Nogales, and all the guests would have their stockings full of things made in Mexico."

In the evenings, in the ranch's small bar before dinner, her grandfather would play the piano by ear, and guests would sing. Clothespins were used as napkin rings at Kenyon. After ten years, guests received a commemorative silver clothespin. For twenty years, the dates and letters "K-R" in solid gold were soldered onto the silver.

"When my grandmother decided to sell in about 1974, I went to Kenyon to see it one more time, because it was always so special to me. It was bittersweet to be there again after so many years." Vera currently lives in New York City and upstate New York, where her husband, an architect, grows grapes and makes wine. "A friend here was a guest at Kenyon. He remembers my grandparents well."

Today, the ranch is geared more toward retreats. Noted for its silence and spirituality, guests come to the ranch from as far away as Kazakhstan to attend retreats. Many comment that crossing over the ranch's cattle guard is like entering another world. "I most enjoy keeping the energy balanced for guests so that the place feels nurturing," says ranch manager Joyce Winough. A Native American sweat lodge, hiking trails, a pool, reflecting pond, hot tub, labyrinth, an art room and massage rooms all await retreaters. Reiki and reflexology treatments heal the body and soul. The ranch's collection of books on Arizona history can be perused at the Tubac Historical Society library.

Charles and Barbara Findeisen, originally from California, have owned the ranch since 2003. Vera stayed again at Kenyon under its present ownership a few years ago and found to her delight that, despite a change in focus, "much of what was very special about Kenyon is still there."

Location: HC 65 Box 278, Tumacacori, AZ 85640, Tel: (520) 398-8073, E-mail: kenyon.ranch@earthlink.net, Website: www.kenyonranch.com

Accommodations: The ranch's seventeen casitas can accommodate up to fifty guests. All-inclusive, custom-made vacation packages are available for spiritual and personal vacations, professional meetings and conventions. Special diets of vegan and gluten free are available.

Chapter 18

Better than Disneyland

Farm Stays in the Land of Sunshine

"City people still need the joys and friendliness of unsophisticated, back to the land vacations."
—Pat Dickerman

Influenced by the local foods movement and, of course, the *City Slickers* movie, there is an increasing interest by the public in farm and ranch vacations across the United States. A "farm stay," i.e. farm vacation, offers accommodations on a working farm or ranch, where working refers to the production of crops or livestock. The term "farm stay" originated in Europe, and an increasing number of Americans are becoming familiar with the travel niche.

An established vacation option in Europe for over thirty years, farm stays have been slower to develop stateside. "I can't say we are less family farm friendly than Europe, but there is no national tourism 'power' to help with the marketing of this 'new' brand. While the USDA may get involved at a small level because it helps its farms, farm stays as a (vacation) model have to carve a niche for themselves state-by-state," explains Scottie Jones, owner of Farm Stay U.S. "There are even those that would argue the government shouldn't assist niche farming at all because it is less efficient than mega farming," says Jones. "Selling produce and/or livestock exclusively can be a difficult way to make a living on a small scale."

On the East Coast, a number of the farms that were familiar with the farm stay concept in Europe decided to try to duplicate the experience for their own benefit. Similarly, on the West Coast, in the 1990s, the UC–Davis Extension Services became interested in helping farms move towards agricultural tourism ventures. "They promoted the development of farm stays through legislation that exempted farms from food-handling requirements if they only served breakfast," says Jones. This paved the way for many farms to open a bed-and-breakfast component to their existing operation. "All the farm stays in between those geographic areas started up mainly on their own without any official body smoothing the way." Given the amount of time needed to plan, build and schedule these operations, farm stays are usually started by those with the "gift for hospitality."

Until recently, few travelers envisioned staying on a farm as a vacation. Prior to World War II, "many people had family farms to return to," observes Jones. Vacationing on a farm simply was not "a novelty" form of tourism…yet. Even after the Second World War, the thought of staying on a farm overnight for rest and relaxation was unappealing to many Americans who preferred to leave the memories of hard, manual labor for the "white collar" workforce.

The focus during this time period was shifting from small farms to a much more industrialized form of agriculture. Small, family farms saw huge declines in the following decades as they faltered to compete. "It has really only been in the last ten to fifteen years that our society has started to question large-scale food production in terms of our health, the health of our animals and the health of the environment," adds Jones. "Now, there's a resurgence among the twenty- to thirty-something generation to recreate that lifestyle and to bring their children up in it as well." Michael McKenzie, owner of the McKenzie Inn Bed and Breakfast, a farm stay in Eloy, Arizona, agrees. "With large family farms slowly selling out to developers the next generation is returning to the roots of their grandparents and rebirthing the old style of community…Younger people are seeking community and a sense of belonging to something greater than themselves."

The rising tide of a younger generation of people interested in farming as a combined career and lifestyle choice is well matched by urbanites who just want to unplug, walk in the woods, sit by a creek and mentally and physically rejuvenate. "Farm stays are a great way to educate urbanites," says Jones, thereby "potentially educating the next generation of farmers."

Just as travelers at the turn of the century wanted to escape an increasingly mechanized society by visiting guest ranches, farm stay vacationers seek to escape their iPhones, iPads and iPods to a more organic world. "It brings one back to a much simpler time in many travelers' eyes," says Jones, who owns and operates her own farm stay in Oregon.

The appeal of the farm stay is obvious. "The different sounds on a farm provide a variety of unexpected opportunities for couples and families to play, talk," comments Jones. "Kids write that this is better than Disneyland, without prompting, in my guestbook." In Southern Arizona, staying on a farm often provides a natural respite from the heat and concrete afflicting most large urban areas in the state.

Aesthetically speaking, "farms and ranches provide green space in often otherwise developed areas," contributes Sarah Potenza of World Wide Opportunities on Organic Farms (WWOOF). "While they may not be a wild ecological environment, they are important open space that is kept from being paved, produces food and provides some wildlife habitat."

This lifestyle has an economic ripple effect. Farm stays contribute economically to their immediate communities in many beneficial ways. When farm overnighters, as opposed to day trippers, spend locally, three times more money has the potential to stay in the immediate community. "If we could redirect the millions that are spent by Arizonan vacationers in Southern California into rural Arizona, the state would recover economically," says Kimber Lanning, executive director of Local First Arizona.

"With the downturn of the economy, farm stay visits can be much more economical than your average vacation and highly educational to boot," says Jim Dumas, owner of RichCrest Farms, a farm stay in Cochise, Arizona.

There are also less quantifiable benefits. Southern Arizona's landscape and ecology is special, and there is no better way to experience this than to stay in one of the more rural parts of the state. "We always come to Southern Arizona in the spring to enjoy blue skies and dry weather, so different from our biome back east, and to see completely different flora and fauna. This area of the west is such an outdoorsy community," comments Cherry Valley Ranch guest Bonnie Halchin.

"The quiet; the hard work that goes into livestock and crop production that translates into those higher prices at the farmer's market; that an egg needs twenty-six days before a chick is hatched; that true free-range chicken eggs have a dark yellow/orange yoke from all the protein they get eating worms and bugs…" These are just a few of the many surprises, discoveries and lessons that guests receive during their farm stay experiences around the country, reports Jones, whose website brings farm stay owners of all kinds together to share, learn and network.

Like classic scenes of life on the farm, the most important preparations for a farm vacation are simple. A few suggestions from farm stay owners are to be "mindful and quiet, listen to instructions, slow down, stay out of the way when necessary, don't expect to be waited on, jump in and help if you want to, don't bring your city expectations to the country and appreciate how most farmers and ranchers consider themselves stewards of the land."

And most importantly bring a hat, sunscreen and boots.

Chapter 19

Agua Linda Farm

Amado

Redefining Glamour

Agua Linda Farm's historic adobe hacienda sits enchantingly along the Santa Cruz River in Amado, only thirty minutes from Tucson. The farm is easily accessed from the highway.

It is lovely and cool here, like wandering into the *Secret Garden*, with its natural shade and large oak trees. The farm's greenery is not forced. It is spring when I visit Agua Linda Farm, and the land is alive with blooming flowers, green grass, insects and baby farm animals. Horses and cattle graze in fields with the Santa Rita Mountains in the background. Little girls in their Easter best chase each other and shriek with laughter. Everyone is here for the Easter egg hunt and celebration. Spanish music floats through the air as children play on tire swings and hunt for Easter eggs and families picnic on the grounds. Hay rides take visitors out into the garlic, onion and tomato crops.

This sixty-three-acre organic vegetable farm has an undeniable glamour. The property's buildings, including a main house, a guest house and a newer pool house, all have an English country feel of more refined days gone by. The history of this place dates back to the 1760s, when the King of Spain granted the Agua Linda area to Torvivio Otero. The property remained in the hands of the Otero family until 1940.

Carlos Ronstadt, Agua Linda's next owner, raised cattle and farmed cotton, alfalfa, corn and barley on the property. Born in Tucson in 1903,

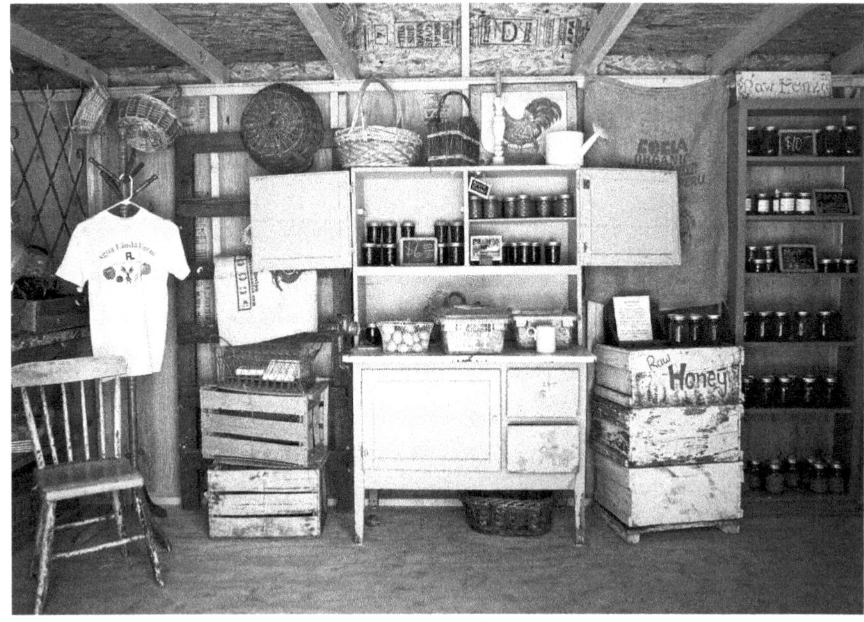

Open year-round, Agua Linda's Farm Store carries fresh, seasonal vegetables from their gardens; raw, local honey; a variety of organic jams and local goods; handmade soaps; and grass-fed, hormone- and antibiotic-free beef. *Courtesy of Amanda Rockafellow Photography.*

Ronstadt became one of Southern Arizona's most prominent cattlemen and businessmen. The Ronstadt family had been active in Arizona ranching since the beginning of the twentieth century and owned a vast spread near the United States border with Mexico. In 1933, after his father's death, Carlos became the head of the family business: the Baboquivari Cattle Company. Baboquivari acquired the Agua Linda properties in 1949 and did extensive development on the land.

In 1957, the one-thousand-acre Agua Linda farming operation was bought by Arthur Loew Jr. from a Hollywood dynasty that included Adolph Zukor, the founder of Paramount Pictures and Arthur's maternal grandfather. Adolph Zukor emigrated from Hungary, and the family lived on the Long Island estate, Pembroke, which has since been bulldozed. Arthur's paternal grandfather, Marcus Loew, founded Metro-Goldwyn-Mayer Studios (MGM) and Loew's Theatres, and his father, Arthur Loew Sr., was president of MGM. Arthur Loew Jr., who had spent much of his childhood in Southern Arizona, loved Arizona and decided to return.

In the 1950s, the opening scene to the classic film *Oklahoma* was filmed in the farm's cornfields, and the place quickly became a kind of Hollywood hide-out for years thereafter. Black-and-white photos of Hollywood stars who visited the farm during its heyday in the 1950s and 1960s, including Natalie Wood, James Dean, Elizabeth Taylor, Paul Newman, Gene Kelly and John Wayne, are displayed in its hacienda.

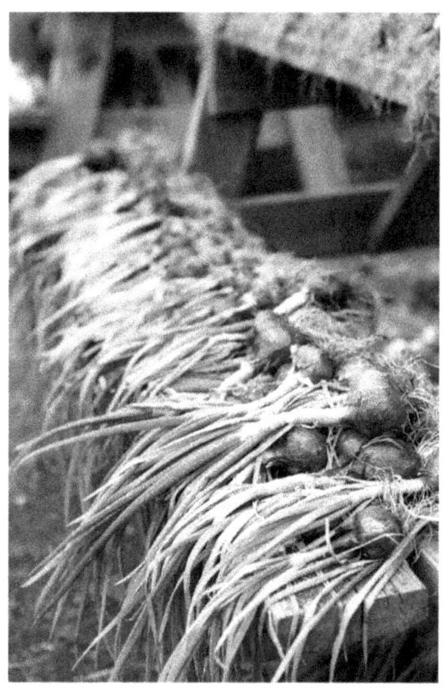

Agua Linda Farm's largest crop is garlic and onion, with pumpkins, melons, lettuce and tomatoes also in the mix. This variety of produce is available at their farm store, local farmers markets and through their CSA (Community Supported Agriculture) program. *Courtesy of Amanda Rockafellow Photography.*

Arthur produced a handful of movies such as *Penelope, The Rack* and *Arena* and traveled back and forth from there to Hollywood. During this time, Agua Linda was used as the feed-lot portion of the cattle business. A feedlot is the final "fattening up" area for the ranches in the greater Altar Valley.

Now, Arthur's son Stewart Loew and his wife, Laurel, are the current owners of the farm. Stewart's mother and former actress Regina created a botanist's wonderland of green lawns, irises, daffodils and more in the ten acres surrounding the main house. "Guests are drawn to Agua Linda more for the general physicality of the place than its crops," Laurel says. "The farm is incredibly beautiful, with adobe buildings, roses and shade. We are in a riparian area that is very lush with lots of trees." The farm also once housed the family show horses that campaigned throughout the 1980s.

In his early years, Stewart saw himself getting into the film industry in some way. Stewart went to New York and worked in sound production and as a fill-in prop boy for David Letterman and *Saturday Night Live* before making the transition to becoming a full-time farmer. Stewart never felt like he was living the "glamorous life" while in New York though. He worked so much and found the experience "rather depressing," explains Laurel. "Stewart feels that the life his family lives at Agua Linda Farm is the real version of glamorous...We enjoy being the magnet to which all our friends and family are drawn and host many gatherings and parties all year. Many of our friends were married here, which caused us to fall into the wedding business...Life doesn't get much better than this!"

When Stewart returned to Arizona, he didn't go right into farming, though. While Laurel was in college, Stewart worked at Film Creations, a studio in Tucson, recording and editing sound for commercials and small productions. He was hired as a sound man for a documentary and traveled to Europe. On that trip, he met Shirley Temple Black, who was ambassador to Czechoslovakia at the time, and she remembered his father fondly. Stewart also helped the Arizona Film Commission, of which his father was a founder, with location searches. "He once met Johnny Depp at the tiny Nogales International Airport and drove him around, but I think his favorite was assisting a shoot with Italian *Elle* magazine with locations for their models!" laughs Laurel.

"When I needed to choose my major in college, Stewart was part-time farming and part-time film industry, so I became a teacher, knowing I could work out in the country or in L.A. if need be," remembers Laurel. "We used to drive around the then-abandoned old feedlot and farm fields that Arthur had sold in our Suzuki Samari with the top off, standing up holding on to the roll bar pretending we were on safari, and Stewart would tell me all the

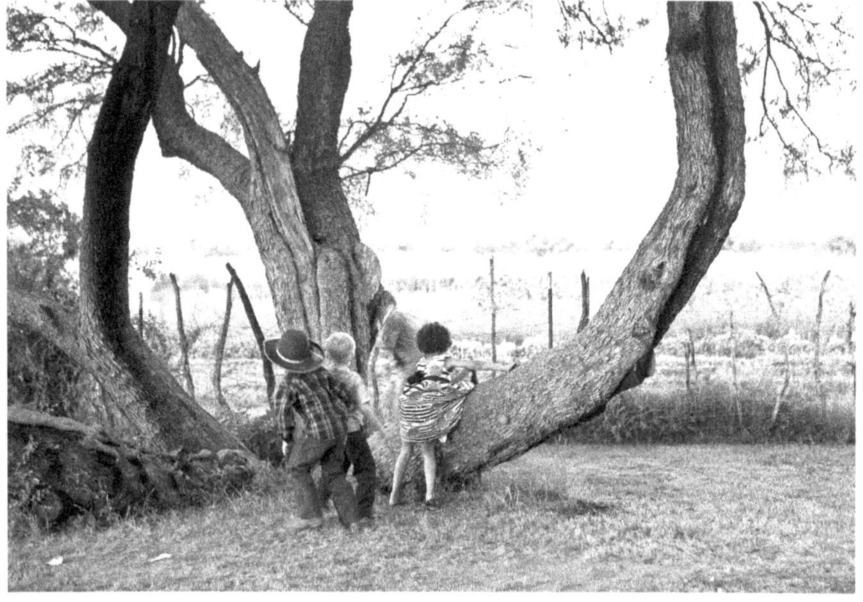

"We have friends over frequently to horseback ride and have 'art nights' when friends bring paints or clay, or even play an instrument in the evening. We host many kid/family days—a giant jumping castle, ponies, baby animals, a hay wagon, endless outdoor games like croquet and horseshoes—we have a blast!" describes Laurel Loew. *Courtesy of Amanda Rockafellow Photography.*

history and how it used to be and how we were going to buy the land back and return it to its former glory. I had come from potato farmers in Maine and had it engrained in my mind that farmers are poor and struggling, so I was slightly appalled at the idea of us being farmers," she says.

What inspired him to make the leap? Stewart says the "arugula started it." His mother, Regina, returned from Europe excited about the greens she had eaten that were so different from the iceberg lettuce so common in this country. The Loews had never farmed before themselves, but Stewart and Regina started planting arugula and lettuce in raised beds in 1993. "I loved the buzz and activity of the operation. I also had come to understand what a precious resource the water right was that my father owned and that if it wasn't utilized in agriculture in some way, it would be lost," explains Stewart.

Currently, most of the labor on the farm is done by volunteers placed by World Wide Opportunities on Organic Farms (WWOOF). The farm draws visitors for horseback riding and scenic hikes, as well as their many harvest festivals, farm tours, farm dinners, field trips and other special events throughout the year. These activities offer the opportunity to learn about Southern Arizona's local agricultural scene and heritage.

"The pumpkin Fall Festival started with one scale hanging from a tree and under five hundred visitors in 2000. We now see something like twelve thousand people throughout the month." The farm's Garlic and Onion Festival, held annually over one weekend, sees about one thousand folks. However, people are still "more interested in the place than what we can grow," says Laurel.

Contact Information: 2643 East Frontage Road, Amado, AZ 85645, Tel: (520) 891-5532, E-mail: stewart@agualindafarm.net, Website: www.agualindafarm.net

Accommodations: Agua Linda's hacienda can be rented for weddings and private events. Accommodation for prospective volunteers on the farm is available throughout an apprenticeship period.

Chapter 20

Across The Creek at Aravaipa Farms—A Country Inn

Winkelman

An Art Lover's Oasis

Across The Creek at Aravaipa Farms–A Country Inn is a farm stay, working orchard and inspiration to countless artists with its intense scenic beauty and palate of color everywhere. The mountains, vegetation and beautiful Southwestern contrasts throughout the farm are noticeable right away.

Located just outside of Winkleman, the farm is next to the Aravaipa Canyon Wilderness Area, home to the second-highest land and mammal diversity in the world. Coyotes, javelina, mountain lions, desert bighorn sheep and coatimundi all utilize Aravaipa Creek, which is one of the few natural, perennial streams left in Arizona.

The Aravaipa Wilderness Preserve also has great significance in the history of Arizona. A variety of diverse groups have left an indelible mark on the land. Its watershed was occupied by hunters and gatherers starting around 9,500 years ago. After seventy Apache women and children were massacred by a group of vigilante Tucson citizens, President Grant sent General George W. Crook to the Arizona Territory (Arizona only became a state in 1912) to implement a federal Indian policy to confine the area's Apaches to a reservation. By the late 1800s, homesteaders arrived in greater numbers to farm and raise cattle and sheep along Aravaipa Creek. Mining spectators opened the area's first mine in 1872. In 1984, Congress officially established the Aravaipa Canyon Wilderness Area to preserve the area's unique heritage and ecology.

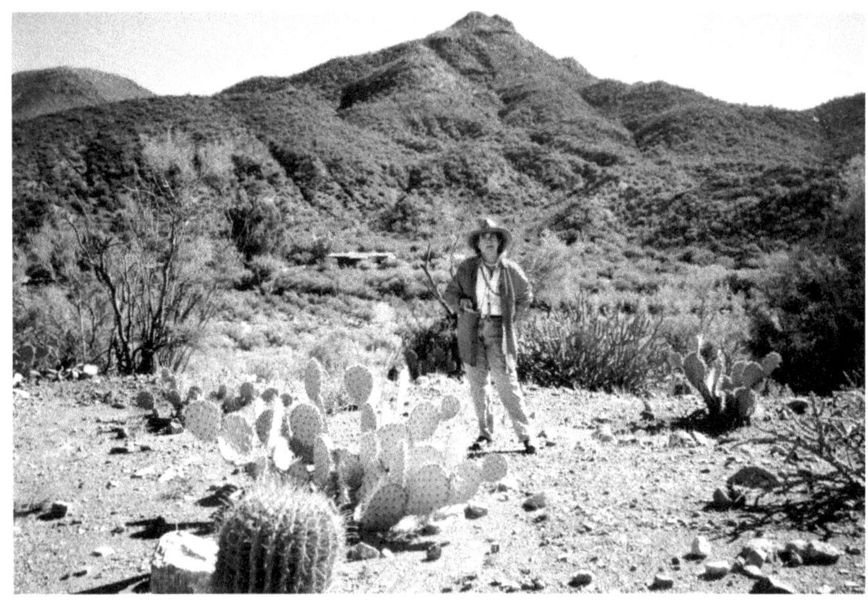

Inspired by the beauty and heritage of the area, Carol Steele opened her farm stay along Aravaipa Creek in 1998. *Courtesy of Carol Steele.*

Situated next to the wilderness preserve, today Aravaipa Canyon is home to roughly forty families, many of whom have been living there for generations. "Everyone who owns property in the canyon is committed to preserving the beauty of the area," says owner Carol Steele, an award-winning chef and retail trendsetter from Scottsdale turned full-time farmer and innkeeper. The Nature Conservancy manages part of the creek that fronts the Aravaipa Farms property.

Irrigated farming had been practiced on Aravaipa Creek for over a century, and organic agriculture has been in place for many of those years. Carol's farm has been an orchard since about 1900 and was once the home of Arabian horses. Starting around 1982, the farm's previous owner planted organic Asian pears, peaches and apricots along the creek.

Inspired by the beauty of the area to open a farm stay, Carol restored and remodeled structures on the property, then added four hundred trees to the orchard, all since becoming its owner in 1998. "I knew that many people come to hike the Canyon and might also enjoy a place to stay and relax."

There is a true sense of "getting away from it all" as you take the highway and head toward the farm, passing desert wildflowers in bloom, snowcapped mountains and bright green riparian areas. The scenery of the upper

Sonoran desert is spectacularly classic Southern Arizona with its rows of giant saguaros. Eventually, the paved road becomes dirt. When you drive through the creek and orchard to reach Carol's property, "you know you have arrived at a unique place," as Carol describes it. The farm sits snugly beneath the Satathite Mountain, separated from the neighboring small farms and ranches.

Carol's home and each secluded guest casita showcase her love of art, rich color and exquisite, eclectic taste. It's little wonder that the place has become a magnet for artists since its beginnings. "They paint the mountains, the creek. There is lots of inspiration if you're an artist," says Carol. The farm hosts famous watercolor artist Ted Nuttal's workshops in the spring and fall.

Professionals, world travelers and even locals come here to unwind, relax and enjoy its peace and stillness. Birding, wildlife viewing, swimming in Carol's enticing pool, jam-making classes and spectacular hiking opportunities only three miles up the creek all await during a stay on the farm. Words really

Aravaipa Farms Morning Banana Cake

Courtesy of Carol Steele

In a large bowl, mix well:
3 ripe bananas, mashed
2½ cups sugar
Add: 2 teaspoons vanilla and 2 large eggs
Add and mix together:
2⅓ cups flour
1 teaspoon baking soda
1 teaspoon salt
Stir together and then add:
1 cup buttermilk
1 cup melted butter
Stir in: 1 cup dried cranberries
1 cup chopped pecans/walnuts

Butter and flour tube pan…10"
Pour in batter and bake 50–60 min. at 350 degrees until pick comes out clean and cake pulls away from sides of pan. Cool, invert cake to remove from pan.

can't do justice to the beauty of this wilderness and farm. A Peruvian couple staying at the farm was moved to tears as they further explored the area. They said they "had never seen anything like it."

Carol's farm quickly becomes special in the lives of all who visit. "Shortly after I had purchased the mountain on 246 acres next to my property, a couple came. They decided to climb the mountain and when they got to the top, the man proposed. They were excited that evening to share with everyone at dinner that they went up the mountain as a couple and came down engaged. Everyone was happy for them," remembers Carol.

Farming still figures prominently in the area's culture and society to this day. Carol continues these traditions on her orchard. Her organic garden and greenhouse supply fresh vegetables, herbs, grapes and berries for guest meals, and a hen house provides fresh eggs for the kitchen and breakfasts. All are grown organically. A self-described steward of her land, Carol's interest in the organic lifestyle stems from "never liking chemicals or technology."

One of the highlights of a stay at the farm is Carol's delicious homemade jams and preserves that are also sold to the local community. Visit the farm during the harvest and jam seasons to watch harvesting activities or participate in a jam-making class taught by Carol. Unique flavors include peach with just a hint of jalapeño pepper. My pre-lunch snack of jam, crackers and fresh fruit from the orchard complement the serenity of an oak tree's shade. A rare leopard frog takes a rest on a casita door handle.

Over the years, Carol's guest books have filled with the fruits of these experiences. "I never realized how much I would enjoy or needed to get away from my hectic life, disconnect from technology and reconnect with myself and loved ones," writes one guest. Creative juices begin to flow. "Guests read, take walks or hike the Canyon, swim in the summer, play games, paint, have wonderful conversations—all the pleasures in life needed to relax and unwind," says Carol.

Carol's family and friends have also taken notice of the uniqueness that is Aravaipa. "We've had some really special times here. Every year, the family gathers for a fabulous time at Thanksgiving. My sister Cathy loves this place as much as I do as a refuge from the city," describes Carol. "For seventeen years, this has been a wonderful place to live. I have thoroughly enjoyed the people and the work." Carol has never looked back on her days working in big city Phoenix.

Contact Information: 89395 E. Aravaipa Road, Winkelman, AZ 85192, Tel: (520) 357-6901, E-mail: carol@aravaipafarms.com, Website: www.aravaipafarms.com

Accommodations: Guest accommodations are private casitas comparable to "living in art" furnished with queen-size beds, walk-in tiled showers, wood-burning fireplaces and kitchen bars. Covered patios, unique water fountains and many Southwestern features invite maximum relaxation. Because of the remoteness of the farm and the typical hiking schedule of the guests, dinner is the first sit-down meal of the day prepared by Carol herself.

Chapter 21

Simpson Hotel

Duncan

The Center of Duncan's Local Foods Movement

A stone's throw away from the Arizona/New Mexico border, Duncan is spread out in the fertile Upper Gila valley with lush farmland and soft rolling mountain vistas surrounding the small town. In this part of the state, locals still wear cowboy hats and boots.

The Simpson Hotel on this cold Thanksgiving evening is a welcoming beacon of history. The building is located on Duncan's Main Street in this small farming community. The hotel is a literary traveler's dream come true with its heritage as an artist's retreat that served as the muse for many a book. The downstairs library holds a collection of books written by authors who have stayed and written here over the years.

The hotel seems to be full of secrets, with its hidden passageways and palpable past. Its resident kittens scurry away as I approach. Art dots the walls. The interior is lush with antique furnishings but most of all—that indescribable antique smell. I breathe it in. My room is the "Sweetheart" room. There is no TV, but a welcoming fireplace complements the chance to catch up on your reading.

Owner Deborah Mendelsohn bought the property in 2006, spending two years renovating and restoring it before opening the full hotel again to the public. Then in 2008, Mendelsohn co-founded the Duncan Farmers' Market to promote local agriculture. "We are more than a B&B. We're the center point of a multifaceted rural development effort."

Wanting to support local growers as much as possible, Deborah grows very few food crops in the hotel garden, mostly basil, tomatoes and garlic. She is also preparing to open a low-water public garden next to the hotel on Main Street, which is drip-irrigated by a rainwater harvesting system. She also has other small gardens in the making that hotel guests will be able to enjoy, including a private shade garden with a fountain behind the hotel's "Old Library" room.

Deborah is very modest about her extraordinary efforts. "I'm not a real grower, just an experimenter," she says. "I started the farmers market because almost all the vegetable stands in the area were closing down due to a range of economic pressures. I couldn't bear to see that, so I helped create a place where very small-scale farmers could gather and sell. I guess the success is that we're still there."

The Simpson Hotel was opened in 1914, known then as the Witt Hotel. The hotel has a significant connection to the Great Orphan Abduction, the historic case that went all the way to the Supreme Court. Jack Simpson, who took over the hotel from Mr. Witt and gave it the name it has today, assisted the sheriff's deputy in preventing riots in nearby Clifton after whites raided the homes of Mexican-American families to carry away newly adopted children who had arrived by train from New York. Mr. Simpson, while helping to prevent serious violence, also made himself a spokesperson for the white mob. "The Mexican-American families ultimately lost their case despite heroic representation by their lawyers," explains Deborah.

Simpson adopted one of the orphans. Anna Simpson grew up at the hotel, and her daughter returned years later to recount its history. Simpson had "Simpson Hotel" painted high on the front and side of the building. The painted lettering is still visible to this day.

The hotel saw its heyday in the 1940s, when it was owned and operated by Second Francese, once the most handsome and eligible bachelor in the Duncan Valley, along with his wife, a graduate of California's Normal School in Los Angeles, who "swept Second off his feet," according to Deborah's sources. "The hotel became a kind of hot spot for the young engineers who came out from Stanford and young schoolteachers from UCLA."

When Interstate 10 to the south was completed in the mid 1960s, Duncan's economy virtually collapsed, and in 1970 or thereabout, the hotel closed. It was purchased and modernized by the Duncan Valley Electric Co-op, which also built a shop and other industrial facilities in the yard in back where once there had been stables, according to old maps.

After a flooding, three successive owners then attempted to restore and reopen the hotel and return it to its former glory without much success. Then Mendelsohn came along. "Fortunately, many of its original features were recoverable, and the place became a kind of family business," says Deborah. Although the hotel in the beginning was "a terrible old wreck," Deborah "fell in love with the area at first sight," she recalls, in the "strange little town of Duncan, here and there were little gems of ingenuity and whimsy."

Growing up in Boston and rural Maine, Deborah dreamed of moving out West. With a degree in comparative religions, she went to Hollywood and produced primetime shows for the networks. She helped support the emergence of non-governmental broadcast news in Russia and Ukraine following the fall of the Soviet Union. "And then I was just ready for a very big change."

Deborah calls the Duncan area the "the deep Southwest." It has remained relatively sheltered from modernization, and it has three cultures that are emblematic of the Southwest: Mexican, Latter-Day Saints and Arkansan-Oklahoman. "It's easy to find people carrying out historic roles in a traditional manner: cattle ranchers, small farmers, gold miners, cowboys, tinkers, quilters and so on," she says.

When Deborah first arrived in Duncan, some townspeople perceived her as "a real estate developer from New York that was going to set up an office in the old hotel to sell property to New Yorkers and ruin the area." She eventually won the confidence of many longtime residents. But that didn't account for the intense interest with which people were peering in the front windows, she says. "Later, I found out the reason for that interest. I bought the building from Ed and Glenna Connolly, who had run a sandwich shop in the front for some years," she remembers.

"Ed was a retired sea captain, originally from the Bronx, and I remember the brogue in his voice when we three first sat down together. He explained, 'You can buy the "motel" but the "motel" comes with Grandma and three goats. Grandma lives in the trailer in back and she'll pay rent, but she has to stay. The goats have to stay too.'"

Deborah agreed to all the Connollys' conditions. But when she arrived to start work, Grandma was gone. "Ed dropped by frequently in those days to see how things were going, and he eventually explained the reason for Grandma's absence: 'I went back and told her, Glenna and I have sold the motel. But you can stay. It's all worked out. You should know though, that the people who are moving in are BOO-dists,'" tells Deborah. "That's how he pronounced it," says Deborah and, "pronunciation aside, it is a fact… He said she gave him an odd look at that moment, but he thought little of it. Within days, the old lady was packing her things, declaring that she had rented a place in the senior housing complex. 'Why are you moving?' Ed demanded. 'We got it all worked out for you to stay!' Her answer: 'I am not living with nudists!'"

Nudists there are not, and the Simpson Hotel is a property that defies a single category—green hotel, bed-and-breakfast and educational garden—but one thing is clear: it truly is the center of its community. Deborah continues to help the Duncan Farmers Market stabilize and spread to other locations in Greenlee County. "We have held lots of public workshops here on gardening, tree care and small business development," she says. "Many local people make use of the public space inside for meetings of various kinds, usually with no charge."

Other unique workshops at the hotel enable guests to learn skills such as canning and drying garden produce. Scenic drives, bird-watching, hot springs and historic-site viewing are a few of the many activities to enjoy in the surrounding countryside. Volunteering at the hotel for a time as part of the World Wide Opportunities on Organic Farms (WWOOF) program is also an option for a more in-depth, comprehensive experience.

Although the hotel uses modern green housekeeping practices, physically, the Simpson really resembles the very prototypical haunted historic hotel that this state is famous for. There is an extreme sense of the past here in everything—furniture, décor, lighting, even toys from another time make it palpable that those who came before are still with us. And a good ghost story is born.

"When I renovated the building, my goal was that when you pass through the heavy, old front doors, you should immediately feel that you have traveled back a century," said Deborah. I think she has succeeded, as I drift off to sleep listening to kittens playing on the stairs.

Contact Information: 116 Main Street, Duncan, AZ 85534, Tel: (928) 359-3590, E-mail: innkeeper@simpsonhotel.com, Website: www.simpsonhotel.com

Accommodations: All guest rooms are furnished with plush bedding, tasteful antiques and private bathrooms that include old-fashioned claw-footed tubs.

Chapter 22

Sojourner's Homestead Bed and Breakfast

McNeal

"Recollection of quality remains long after the price is forgotten."
—E.C. Simmons

Gateway to McNeal's Serengeti

The small town of McNeal in Cochise County is off the tourist radar. Those who find Tucson "ruined" retreat to this place. The town's spirit of individuality comes through right away. McNeal is home to one of the largest Quaker communities in this country, to the McNeal Mercantile Company Store opened in 1919 and, more recently, a quaint Arizona farm stay named Sojourner's Homestead Bed and Breakfast.

"Hospitality is a gift often not nurtured," writes one guest in their guestbook. At this small historic farmhouse a few minutes from central McNeal, there is immense hospitality everywhere. Homemade cookies hot from the oven, plush bathrobes to slip into after a long drive and roaring wood-burning fires.

The place is also much lovelier than its website reveals. Architecturally, the farm resembles a farm stay of Pennsylvania's Amish country. The house's charm is in the details—quilts, kerosene lamps, framed wedding handkerchiefs from the 1950s and crystal lamps. These relics were all donated by locals—Sojourner's is the creation of the entire McNeal community.

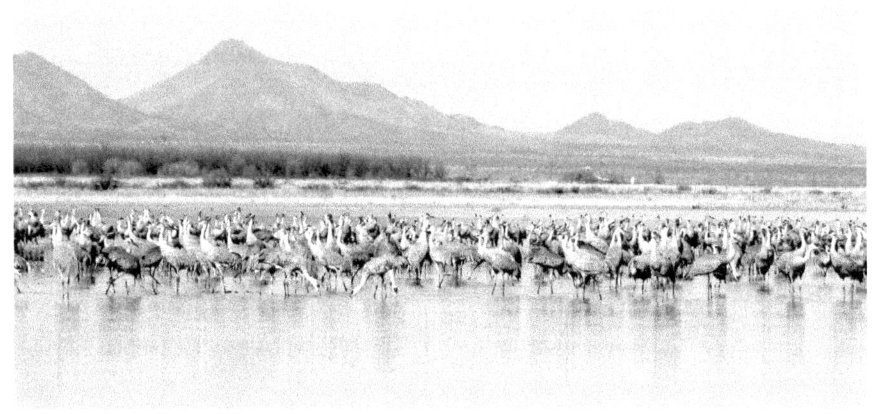

Sojourner's Homestead Bed and Breakfast is only two miles from Whitewater Draw Wildlife Area, Arizona's best place to view sandhill cranes. *Courtesy of Ann Haver Allen.*

Everywhere, you experience the sky—at night, it's black and bright with a million stars, and a large spa in the old laborers quarters overlooks the setting sun. Sandhill cranes fly above to their winter nesting grounds at Whitewater Draw, and a coyote yips as the sun rises.

Over a hundred years ago, the Sulphur Springs valley where the farm is located was a vast open range for cattle and horses dotted with a few small ranches. By 1918, nearly all the land had been homesteaded or used for cattle ranching.

Originally called the Franklin farm, Sojourner's Homestead stands with the shadows of the Swisshelm Mountains as its backdrop. The land is flat all around, and dramatic sunrises and sunsets can be viewed from each guest bedroom porch.

Built in the 1930s, the farm grew everything from potatoes to peanuts and sold its crops to the local community for a number of years. The street running by the farm was named for the farm's original owners.

Drought and crop failure came, and the farm sat vacant until the second owners, farmers Annie and Bernie Carper, bought it. Then the house was abandoned for a time.

The farm's current owners are Terri and Ken Allen. Talkative and down-to-earth, Terri toured different farms and bed-and-breakfasts before purchasing Sojourner's in 2001. Originally from New Jersey, they saw owning a farm bed-and-breakfast way out in southeastern Arizona as a chance to "get out of the rat race." They never looked back. "I love the open spaces of Cochise County. Here, you know your neighbors but still have space." Since "to sojourn means to travel, to journey," the farm was renamed accordingly.

The décor of the house reflects their Northeastern roots—old-fashioned fireplaces, dial telephones, lace curtains. A self-taught carpenter, Terri saw the dilapidated house as a place to practice her skills. "Building is a form of relaxation for me." Although Terri and her husband made many structural changes and modern improvements to the house before opening to guests, they managed to expertly maintain the historic integrity of the farm.

"The house has preserved a lot of the old," says Terri, pointing out the original linoleum floor and old icebox full of mason jars. People tell her to expand, but she likes to be able to cater to each guest on a more personal level. "A large staff has no interest in the place."

The farm's small garden produces tomatoes, peppers and green beans that go right into guests' country meals. Free-range, organic eggs are sold on-site to travelers, guests and McNeal residents. Enjoy picking grapes, apples and pecans around the property. Simple pleasures seem to reign here. Try hand-pumping water from an old well, toasting marshmallows or calling chickens.

The forty-acre countryside surrounding the farm is picturesque, silent and serene. Catch a glimpse of bobcats (they love the hen house), javelina, mule deer, coyotes, jack rabbits or other native desert animals on an evening stroll.

Whitewater Draw Wildlife Area, a significant but little-known bird sanctuary is two miles from the farm. Sandhill Cranes fly by the thousands to their nesting grounds at Whitewater Draw. It's an incredible spectacle to see as many as twenty thousand birds flying low overhead in a kind of an avian safari experience.

Although few in Arizona have heard of this quirky farm stay that's a little off the tourist track in Cochise County, guests manage to "flock" to the farm from all over the world—Israel, Italy, Germany. "We enjoy the interaction with people. This is not just a job, it's our home. A farm bed-and-breakfast offers guests more personality and a chance to learn about the local community." Guests become like family to Terri, and Sojourner's enjoys "a lot of mutual community cooperation, respect and camaraderie from the community."

Contact Info: 8832 Central Highway, McNeal, AZ 85617, Tel: (520) 642-3579, Website: www.sojournershomestead.com

Accommodations: Sojourner's themed guest rooms (Green, Garden and Burgundy) come with individualistic amenities, such as a private library, covered deck, an old-fashioned tub and a full country, home-cooked breakfast each morning.

Chapter 23

RichCrest Farms

Cochise

A Serene Place to Harvest the Seasons

RichCrest Farms is not officially certified organic. But does it really even matter? This farm does "things simply, before the advent of chemicals, and we are more organic than most 'certified' organic farms," opinions its owners, Jim Dumas, Tom Richardson and Etta Sechrest. Their lack of officialdom does not impact the taste of some of the most flavorful salsa I have ever had, made with produce grown right on this land.

Located in small-town Cochise, RichCrest is beautiful, a kind of "oasis in the desert, with an intangibly good vibe," comments one guest. Because of this lushness, a great diversity of bird life stops to use the farm as a water source. Buffer zones for wildlife border the property. "We want to provide that habitat for wildlife, they're beneficial," Jim says. Lately, a kind of "Peter Rabbit-esque" creature has been frequenting their garden, causing more than a little mischief. However, the farm owners do not feel the need to control "pests." Predators keep the insect population under control, and the way they see it, "commercial farming is like getting caught in a death spiral for a farmer."

"There has really been a change in the way organic farming is done," observes Jim. "Back in the '70s, it was about stewardship of the land, making the land better than it was. Now, organic farming is more health oriented." The owners of the farm hope to return the farm to that original purpose as well as to incorporate the old with the new. Richardson and Sechrest founded

RichCrest Farms, a combination of both their names, along with Dumas in 1996 as an extension of their specialty food business. "Etta can plant a rock and make it grow, and there's nothing else I'd rather be doing," says Jim, a self-described minimalist. In a state where many crops are inedible, hers is truly a talent to be treasured.

Under the motto "We Harvest the Seasons," RichCrest sells all their produce and products locally. Tucson, Southern Arizona's major city, started as a traditional ranching community that became cosmopolitan only recently. Tucsonians, known for their interest in local foods, the environment and native plants, make up the majority of guests at the farm. Catch RichCrest at Maynards Market, Jesse Owens Park, Plaza Palomino or St. Phillips Plaza in Tucson.

An old Butterfield Stagecoach route runs right through the farm, and relics of its days were recently uncovered by the Willcox Historical Society. In 1975, Doug and Evelyn Corron founded the farm. They were beekeepers who also grew fruit trees, garlic and artichokes and sold earthworms. Evelyn

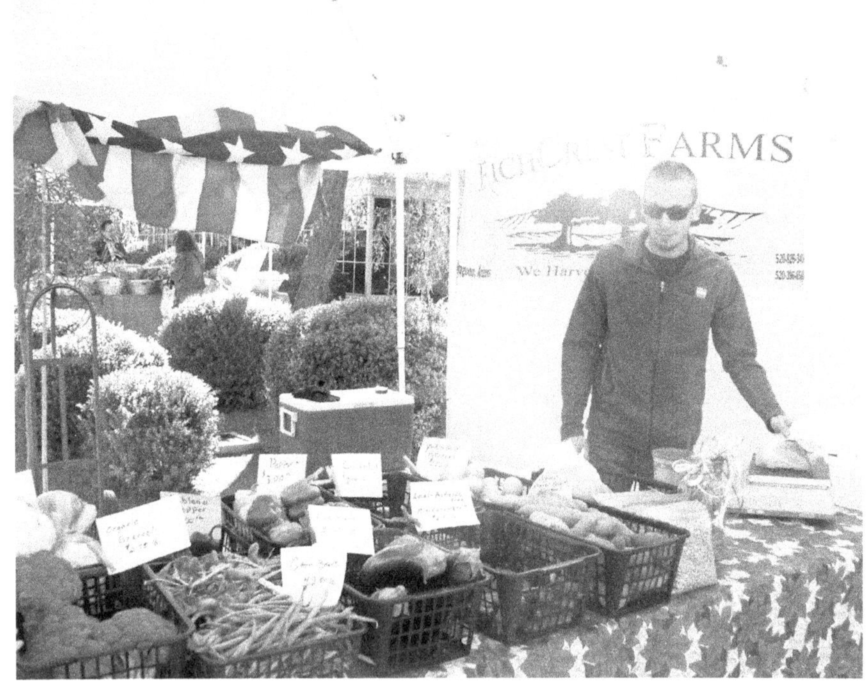

Best known for their garlic, onions and salsa, RichCrest Farms sells its locally grown organic produce and specialty food products at farmer's markets throughout Tucson.

RichCrest's Green Chili Pie

"This recipe was handed down from my mother. She figured it was simple enough that us boys couldn't mess it up too bad," says Jim.

18 corn tortillas
8 medium-sized, roasted (or out of a can) whole Anaheim peppers
2–3 cups chopped onion
2 twelve-ounce cans of cream of mushroom soup
3–4 cups grated white and or yellow cheese

Lightly grease the bottom of a larger size casserole pan (with olive oil or your choice) and lay down 6 uncooked corn tortillas to sufficiently cover the bottom of the pan.

With a spoon liberally spread soup over the top of the tortillas (just like putting icing on a cake), then add a layer of peppers, onions and cheese, saving enough of the ingredients to form a second layer in the pan. Lay another layer of tortillas over the first one and finish it off in the same fashion. The third and last layer will consist of tortillas and cheese only.

Cover pan and bake at 350 degrees for 40 minutes, uncover and bake another 10 minutes to brown the top. Serving size: six.

was beloved in the community around RichCrest and is still remembered to this day. The dirt road on which the farm is located was named after her and the couple's good deeds.

A born and bred Southern Arizonan, Jim's mother was an avid gardener. His father worked at a pipe and steel business that catered to local farms and ranches. His great-uncle was the mayor of nearby Benson. He has lived in the Sulphur Springs Valley for the past forty-seven years. Farming was a natural fit for Jim with the "peace of mind" it provides.

Despite a lack of mechanization and part-time sporadic labor, over the years, RichCrest has grown in capacity. Once a field of under two acres, the farm has grown to eighteen total acres of land, with six acres under production. Garlic is grown "by popular demand," in addition to okra, black-eyed peas and many other vegetables, with an orchard as a coming attraction.

In RichCrest's community garden, members can plant their own crops, use RichCrest's water and land and in exchange, help on the farm for a few hours. "This was a way to solve our labor shortage," says Jim. "Our challenge is to attract more young people into farming, to get young people back on the land." U-pick veggies are available from June through October each year. They have a side business selling and making local products, such as truly delicious sauces, salsas and relish. RichCrest Farms does not keep regular "open to the public" business hours, as they put it. Being open by appointment/reservation allows them to "spend more time with the people who do come out," says Jim.

Over the years, guests have faithfully returned to this diamond-in-the-rough farm after their first stay. Overnighters can participate in harvesting and planting parties to help "get the fields ready." Volunteers always take away some of the harvest or are fed on-site. An annual garlic planting event in October allows visitors to spend the day on the farm working side by side with farmers. Guests learn proper planting techniques, care of crops and harvesting tips.

Other farm highlights include hayrides, farm tours and the chance to take on-site sustainable living classes taught by local experts. Participants learn about a wide range of topics: gardening, water harvesting, animal husbandry, canning, solar and alternative energy and more. The RichCrest Country Store features gourmet homemade jams, jellies, salsa, relish, vinaigrette, grape seed oil and other local products available for purchase.

With its eye on a bright future, RichCrest is slowly building its infrastructure. Most notably, locals have dedicated time and talent to the community adobe building/art project that provides work facilities for volunteers. Marilyn Zwack, a local artist and friend of the farm, designed the building with "lots of positive energy."

"Our guests are always surprised by how much work farming really is," says Jim. Most guests and visitors leave with "a new appreciation for food cultivation."

Contact Information: 2768 N. Evelyn Lane, Cochise, AZ 85606, Tel: (520) 826-3434, E-mail: searchthegardens@hotmail.com, Website: www.richcrestfarms.blogspot.com

Accommodations: RichCrest Farms is open year-round to the public interested in a real small-farm experience. The rustic Sunset Bunkhouse comes with a fully equipped kitchen and private bath but no television or internet access. Camping and picnic areas are also available.

Chapter 24

McKenzie Inn Bed and Breakfast

Eloy

It takes someone to be the First

It is a cool day in Arizona. Fluffy white clouds dot the sky. This is what Arizona is all about—"cattle and cotton," says Michael McKenzie, co-owner of the McKenzie Inn Bed and Breakfast on Lucky Nickel Ranch, a fairly new farm stay just outside Eloy. "Lucky Nickel" is his daughter's nickname. In November, his farm is just beginning its planting season.

The inn is a one-story, red-roofed home surrounded by peaceful farmland. Michael made an architectural design of his and his wife, Pani's, dream house, consulted an architect, rented a tractor and the McKenzie Inn Bed and Breakfast was soon built. One of its unique features is a kind of speakeasy on either side of the front double door of the McKenzie Inn. The home features an architectural design called a "Greek cross"—a very long, left-to-right hallway with the entry and dining area serving as a high cross.

Inside, black-and-white and color photographs, as well as their daughter's original artwork, give the house a sense of family importance and history amid the impeccable interior polish. Each guest room in the house is named after a family member. A wood burning stove, popcorn popper, Singer sewing machine and an antique bed and dresser make my room both old-fashioned and new.

Cotton and alfalfa fields and mountain vistas surround the property. There are a few neighboring ranches closer to the main road. Birds chirp

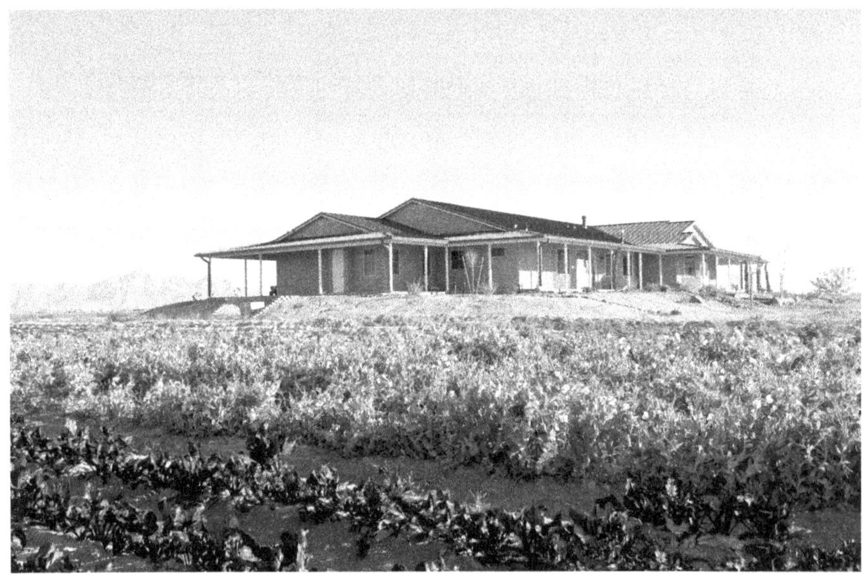

The McKenzie Inn Bed and Breakfast on Lucky Nickel Ranch. This trailblazing farm stay, surrounded by cotton and alfalfa fields and mountain vistas, is a "labor of love" for its owners. *Courtesy of Rosemary Woods.*

or pick at insects on the ground. You can hear a train in the distance. Farm sounds and smells permeate the air—horse hoofbeats, hay and Bermuda grass. The farm is across from Skydive Arizona, the number one skydiving site in the country. Today, the silhouetted figures of sky divers and hot air balloons can be seen above.

The McKenzie Inn Bed and Breakfast is part ranch, part farm and part wildlife haven. It's owners also do not fit into one category. Michael was an Intel technician turned full-time organic farmer. His wife, Pani, worked as an insurance agent, then "got the calling" to become a school teacher and part-time innkeeper. They are both very ambitious. "Small farms must diversify in order to stay afloat," says Michael.

After being laid off from an aircraft manufacturing job in California, Mike vowed "to live more simply and to never put work before family." After seeing the Christian movie *The Ride*, he began to picture himself as a farmer and rancher and immediately started praying and looking at different properties in Arizona to open his own farm. The only piece of land not getting a definite "no" from God, he says, was the farmland outside of Eloy.

Then called Sunland Gin Farms, Michael bought the property in 1997 with the help of partner Tom Fuhrman and started to build Lucky Nickel

Ranch from the ground up. The McKenzies were the first to buy land on the then-sprawling farm of forty-two acres. Other families soon followed and built their own homesteads on different parcels of the original farm. The family started the inn portion of their farm stay endeavor in 2008.

The McKenzies' family history is as inspiring as the development of their business. They are completely self-taught. Michael's family came out West to California from Oklahoma and Texas (where many still live today) during the early 1940s. Pani grew up in Greece and went to high school in the United States. An ESL student, her father once hilariously told her math teacher, "She's not stupid; she's Greek. We were forgetting things before you had even learned them." Pani has inherited his outspoken nature.

Before he was able to farm successfully, Michael experienced his fair share of learning curves. "It took me so long to learn how to irrigate, and that was a disaster my first year. I was shoveling dirt everywhere just to keep the water on my crops."

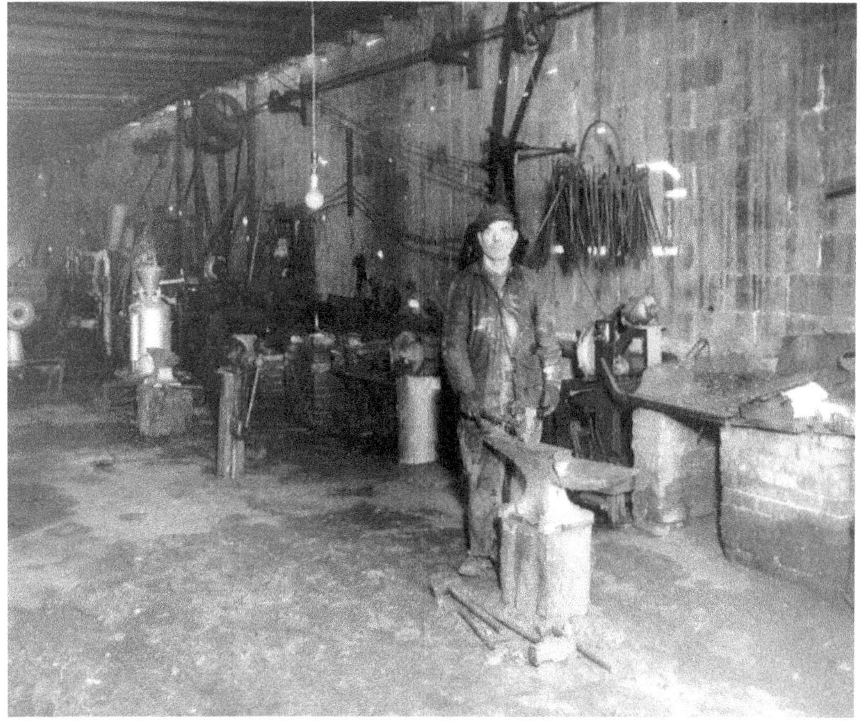

A self-taught farmer, Michael McKenzie's grandparents originally migrated west from Oklahoma and Texas during the early 1940s to work in the grape vineyards of Earlimart, California. *Courtesy of the McKenzie Inn Bed and Breakfast.*

"I really thought I could make more money working for myself as a farmer." He quickly learned that few banks make agricultural loans to new farmers, and "the system seems set up for large farms or farms that have been in business a long time," says Michael. Most large-scale farms are second and third generation. Those "beginning" farmers are already set up for it when the time comes—the land, the equipment and the connections are already in place.

So how can a city slicker with a dream start his own farm of any size? Michael financed his small-farm dream from his own funds and clever bargain shopping (a $7,500 1980 tractor is apparently a bargain), trial and error and making the right connections over time. "Farmers don't do anything really quickly," he observes. "They'll think about it, chew on it, then three months later, do it."

With these truths in mind, the McKenzie farm's success is even more amazing. They have supplied Whole Foods in Scottsdale, Rio Salado College's Café Rio as well as Bon Appetite Cafés in Chandler. Local chefs buy directly from the farm. They were the only small farm to be invited to a Whole Foods–sponsored barbecue in Scottsdale attended by John McCain.

Not satisfied to merely break even on his investment, through farming, Michael hopes to "employ and improve life by creating decent jobs for people interested in small scale farming." The McKenzies run a Saturday farmers' market on site at Lucky Nickel Ranch. Events such as cooking demonstrations by chefs using the farm's specialty produce, a mindful-eating workshop and retreat, a corn maze and pumpkin patch are some of the events going on throughout the year at the farm.

Soon the farm hopes to market its produce to the Olive Garden and Red Lobster, among other establishments. A greenhouse in progress will supply fresh herbs and tomatoes, and their own brand of grass-fed, organic beef is in the works.

"With every opportunity that diversifying offers a farm—having 'u-pick' veggies, a bed-and-breakfast, selling to local restaurants—another challenge is presented," says Michael. "You can have a few good years, then a bad year, then another good year," agrees Pani.

The property has more than produce and horses; it has wildlife. The farm's burrowing owls nest a little beyond the Bermuda grass field. A coyote basks in the sun with a new friend—a buzzard. Bats roost in the toolshed. With a grant from the National Resources Conservation Service, part of his land is devoted to a wildlife habitat. Muslim Uzebek refugees who worked on his farm a few years ago would "talk to the rabbits" to form an agreement to

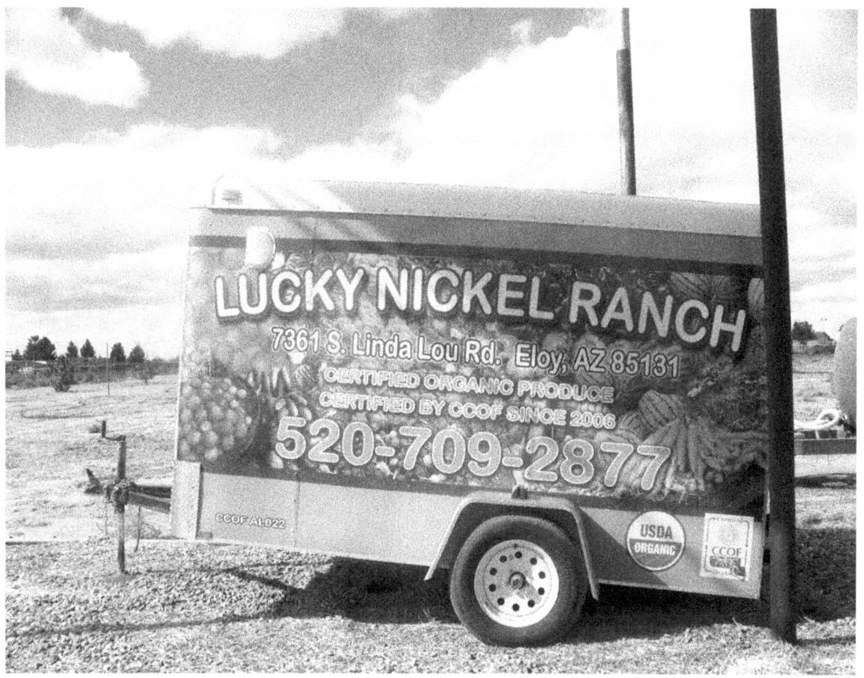

The McKenzies operate a farmers' market on Lucky Nickel Ranch, selling their organic, seasonal produce. "We started farming with the vision of impacting our community first and then farming around that model. It is an old-school way of doing things, returning to the roots of our grandparents and that small farm generation," explains owner Michael McKenzie.

only eat a little of their crops—or else. So far, it's working, and pest control is not a problem.

One professor comments that "Michael is actually doing what academic types are only writing about." Michael is optimistic that in the near future, people will be "begging" to work the land, but he admits that it "takes someone to be the first…We are a small farm and can't compete with large-scale agriculture, but we feel we fulfill a different niche in the community anyway," he says.

The farm is even more beautiful at night. The land is so silent. It feels like you've left all civilization behind even though the farm is about equidistant from Phoenix and Tucson. Rabbits dash here and there. A hawk eyes them from his perch. The distant mountain glows with purples and pinks. All I hear is the wind. I search for barn owls that nest in the hay stacks, but no luck. I decide to feed Blaze, Lauren McKenzie Cross's quarter horse/

appaloosa. The county life, with its unique pace and feel, slowly seeps in. It takes a while to slow down. After a day, this farm vacation begins to work its therapeutic magic.

Morning's misty air smells of fresh hay. I perk up with some strong hot coffee and have Pani's freshly made organic spinach spanakopita (inspired by her heritage, with spinach right off their farm). After breakfast, a quick trip to the burrowing owl sanctuary on the other side of the property yields a few sightings of the owls and jack rabbits.

These past few years, the farm has been a "learning lab" for the local community on organic agriculture. Since they started over a decade ago, organic farming and gardening has since spread like wildfire around the country. Although Marshall Trimble writes that historically, the "life of a farmer has never been the subject of romantic lore," Michael's farm seems to be turning that perception on its head.

Contact Information: 7361 S. Linda Lou Rd. Eloy, AZ 85131, Tel: (520) 709-2877, E-mail: info@luckynickelranch.com , Website: www.luckynickelranch.com

Accommodations: The McKenzie Inn features three rooms each with their own special charm and amenities to enjoy reflecting the character of the person who inspired its decor. An excellent breakfast menu includes selections such as Mediterranean quiche, casserole, yogurt pancakes and raw honey.

Chapter 25

Almuniya de los Zopilotes

Patagonia

Bringing Food Security to Santa Cruz County

Opened in 2010, new farm stay Almuniya de los Zopilotes or Private Experimental Farm of the Turkey Vultures is making local history as the first of its kind in the state. Owned by award-winning Lebanese American author and conservationist Gary Nabhan and his wife, herbalist and medical anthropologist Laurie, the farm is located just outside of Gary's pedestrian-friendly hometown of Patagonia (Spanish for "big foot") in Santa Cruz County.

Agriculture has existed in Santa Cruz County for over a thousand years. The area is known for its corns, beans and squash varieties that are still being grown and consumed in much the same ways that they have been since the ancient beginnings of this "phenomenal" tradition of farming. Half of what is currently grown on Gary's farm was planted three hundred years ago in the area.

Gary's aesthetically pleasing farm itself is modeled after the Almuniya, or experimental farms, orchards and gardens used for research and pleasure operating in Spain a thousand years ago. *Ancient Agriculture,* written by a friend of Gary, inspired him to start his own farm with those purposes and to further contribute to and promote food diversity and security in Arizona's borderlands. After visiting Andalucia's Almuniya, it further crystalized for Gary and Laurie that this is just what they always wanted to do—to not just exchange seeds but the stories behind the foods and recipes as well. The Turkey Vultures portion of the farm's name is a nod to Gary's sense

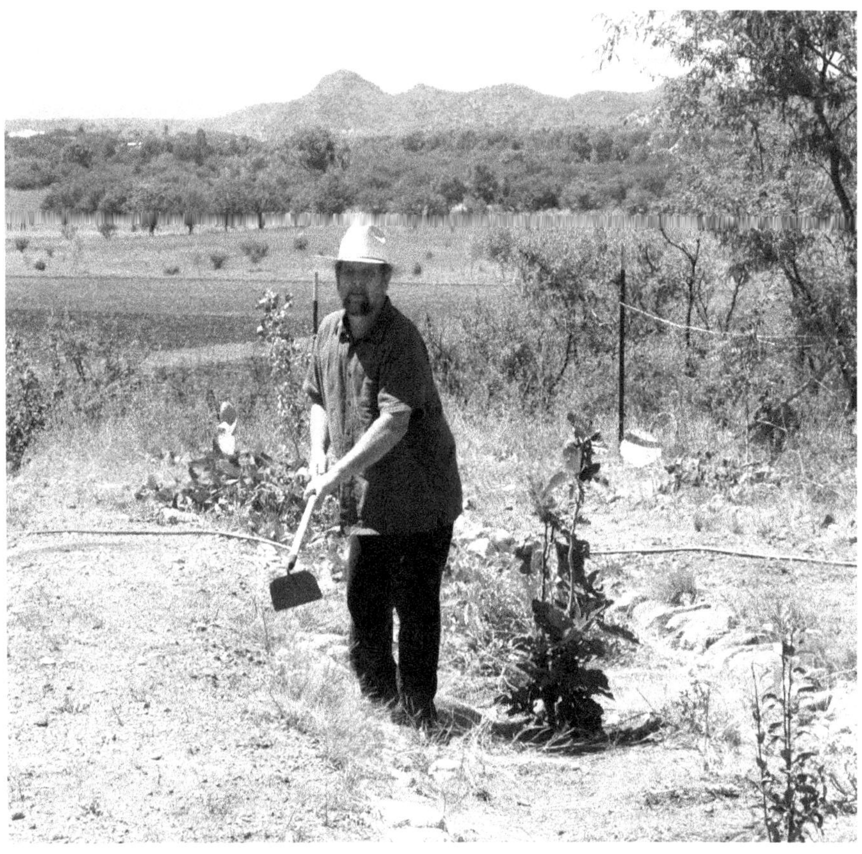

Almuniya de los Zopilotes owner and naturalist Gary Paul Nabhan tills the farm's rich soil in the foothills of the Santa Rita Mountains. Healthy soil equals healthy plants and seeds. *Courtesy of Tim Tracy.*

of black humor: "whatever doesn't make it through the experiment will be eaten by the area's vultures," he says mischievously. An example of some of the farm's recent experiments in sustainable agriculture are optimal pairings such as using a native desert tree to grow sun-sensitive plants under that tree's shade.

The lovely five-and-a-half-acre farm includes views of Red Mountain, wraparound gardens and a permaculture-designed orchard that grows one of the largest selections of Mission-era (brought to Arizona by Spanish missionaries) heritage fruits, nuts, beans, herbs, asparagus, olives, prickly pears and chili peppers in the southwest. Just what doesn't the farm grow? Bee houses provide shelter for these pollinators. Gary's aim is to make this the most diverse farm in Arizona in terms of the number of varieties that are grown here. A combination of rainwater harvesting techniques from desert cultures around the world are used to grow all crops. Terraces catch and conserve this water. "Guests can learn to use these techniques in their own gardens," says Gary, who believes food is "just too important" to leave completely up to the government. The family hopes to sell "value added" products (jams, jellies, pestos) from their canning kitchen throughout Santa Cruz County in the coming years. Local bakers and chefs conduct workshops in the farm's kitchen on how to cook using heritage foods. The family serves its guests one communal meal a day and a separate private meal. Gather as much as you can eat from the farm to snack on in the meantime.

Almuniya de los Zopilotes guests can participate in a variety of activities—such as gardening, swimming, fishing, ecological restoration and terrace building—during their farm stay. Off the farm, enjoy the area's world-class bird watching or vineyard touring in nearby Elgin. Get caffeinated in Patagonia's coffeehouse Gathering Grounds with its laid-

Sonoran Desert Hummus

Courtesy of Almuniya de los Zopilotes

1 pound Sonoran white tepary beans, boiled until tender and drained
¼ cup Mexican lime juice
4 tablespoons sesame paste
½ teaspoon sea salt
2 tablespoons Mission olive oil
2 tablespoons ground lemonadeberry seeds
1 teaspoon Mexican oregano leaves or cumin seed
1 chiltepin

After completing cooking of the tepary beans, drain then put in a blender or in a large molcajete grinding stone. Add lime juice, sesame paste, olive oil, salt, chiltepin and the oregano leaves or cumin. Blend until whipped into a frothy paste. Refrigerate until cold, then place in a ceramic bowl and sprinkle lemonadeberry—an American sumac—in a spiral pattern over the hummus. Garnish the edge with pomegranate seeds. Serves six.

back artsy atmosphere or relax with one of Gary's books near the farm's pond or nursery.

As an added bonus, the farm overlooks the sixty-acre Native Seeds/SEARCH farm that Gary helped found. The nonprofit conservation organization and seed company promotes regionally adapted seed diversity and has influenced the return of heritage farming in the region since its beginnings in 1982. At the Native Seeds farm, Gary's guests can also volunteer in harvesting, hand pollination and planting.

More than fifteen miles away from any full-service supermarket or chain grocery store such as Fry's or Safeway, Gary sees the lack of this service in his community as a major selling point for future guests—visitors to the area can fully experience the variety of local fruits, vegetables and other agricultural products at his farm and in the unique small grocery stores and cafes around town. The farm demonstrates that locals do not have to rely exclusively on food purchased at larger chain stores—it is possible to grow and provide for your own dietary needs using varieties of seeds adapted to the region's climate and environmental conditions.

It is amid this spirit of sustainable agriculture that a stay at Almuniya comes alive with meaning and depth. A research scientist at the Southwest Center for the University of Arizona, Gary mentors interested students and guests in a wide range of subjects, such as desert water-harvesting, seed-saving or climate-friendly farming techniques while they enjoy the many recreational opportunities of the area.

A stay at this farm is not only a unique vacation experience but a lifelong learning experience to boot. Clearly, a great deal of healthy food can be grown easily and sustainably on a relatively small amount of land. I hungrily contemplate a delectable artichoke ripe for a salad. Ancient agricultural traditions of desert dwellers around the world are alive and well in Southern Arizona.

Contact Info: E-mail: gpnabhan@email.arizona.edu, Website: http://garynabhan.com

Accommodations: Almuniya de los Zopilotes's two-room guest house can be reserved for vacation, work and/or study stays All training and volunteer opportunities can be customized according to individual guest interests and needs.

Bibliography

Allen, Marchella Y. "Voices of the Valley." *Tubac Historical Society oral history project.* 18 March 1986: 2–7.

Allen, Paul L. "A People Place." *Tucson Citizen*, 4 April 1992.

Allen, Terri. Personal interview. 27 December 2011.

"Aravaipa Canyon Wilderness Area." U.S. Department of the Interior Bureau of Land Management. 12 September 2011.

Arizona Game and Fish Department. "Whitewater Draw Wildlife Area." 2009.

Arnold, Oren. *Arizona Under the Sun.* Freeport, ME: Bond Wheelwright Co., 1968.

Bearden, Michelle. "Center Short on Residents, Long on Causes." *Phoenix Gazette*, 22 April 1989.

Bernstein, Joel H. *Families that Take in Friends: An Informal History of Dude Ranching.* Stevensville, MT: Stoneydale Press Publishing Company, 1982.

Bingmann, Melissa. "Prep School Cowboys: Arizona Ranch Schools and the Images of the Mythic West." *Journal of Arizona History* 43 (2002): 205.

Bishop Jr., James, and Bernie Blake. "The Arizona Rancher: An Endangered Species?" *Phoenix*. June 1992: 57–65.

Bodfield Bloom, Rhonda. "Historic Ranch Revived as B&B—A Dream Come True for Couple." *Arizona Daily Star*, 21 April 2004: H2.

Brookens, Suzanne, and Roger Evans. *Rancho De La Osa.* Alexandria, VA: n.p. 1983. Print.

Burnham, David. *Winter in the Sun*. New York: Charles Scribner's Sons. 1937. Print.

Cardon, Charlotte M., and Ernest Cabat. *Life on the Tanque Verde.* Tucson, AZ: Cabat Studio Publications, 1983.

Coppola, Manuel. "State, City Considering Proposal Making Park at Guevavi Ranch." *Nogales International.* 3 October 1990: 1.

Cosulich, Bernice. "What Lies East of Rincons." *Arizona Daily Star*. 3 February 1929: 10.

Cote, Bob. Personal interview. 25 March 2012.

Courtland Arizonan. "Considerable Activity Around McNeal." 26 January 1918.

Daniel, Diane. *Farm Fresh North Carolina.* Chapel Hill: University of North Carolina Press. 2011. Print.

Darimont, Adrian. "Re: Ranch Vacations Book." E-mail to the author. 26 March 2012.

Day O'Connor, Sandra, H. Alan Day. *Lazy B: Growing Up on a Cattle Ranch in the American Southwest.* New York: Random House Inc., 2002.

Dickerman, Pat. *Farm, Ranch and Country Vacations in America.* Scottsdale, AZ: Adventure Guide, Inc., 1995.

Dumas, Jim. Personal interview. 29 December 2011.

———. "Re: Farm Vacations Book." E-mail to the author. 29 January 2012.

Egloff, Fred R. "Circle Z Ranch and the Sonoita Valley." Keynote speech printed in *The Westerners Brand Book*. Prudential Building, Chicago, IL. 30 January 1978. Keynote Speech.

"Employment Projections: 2010–2020 Summary." U.S. Department of Labor Bureau of Labor Statistics. 1 February 2012.

Flood, Elizabeth Clair. *Old-Time Dude Ranches Out West*. Salt Lake City, UT: Gibbs-Smith Publisher, 1995.

Fox McGrew, Joy. "Re: Rancho Linda Vista." E-mail to the author. 20 April 2012.

Franklin, Sidney. Personal interview. 9 January 2012 and 6 April 2012.

Gustafson, Gus. "Re: Farm and Ranch Vacations Book." E-mail to the author. 1 April 2011.

Hadley, Diana, Peter Warshall and Don Bufkin. *Environmental Change in Aravaipa 1870–1970. An Ethnoecological Survey.* Phoenix: Arizona State Office of the Bureau of Land Management, 1991.

Harrison, Sherry. Personal interview. 28 November 2011.

———. "Re: Farm and Ranch Vacations Book." E-mail to the author. 3 January 2011.

Hess, Alan. *Rancho Deluxe: Rustic Dreams and Real Western living.* San Francisco: Chronicle Books, 2000. Hillinger, Charles. "Rancher Wants Mission 'in his Backyard' to be Preserved." *Los Angeles Times*, 12 February 1986.

Hodson, Colleen. "Re: Book on Farm and Ranch Vacations." E-mail to the author. 17 January 2012.

———. "Re: Guest Ranching History." E-mail to the author. 22 November 2011.

Holnback, Sharon. Personal interview. 9 September 2011.

———. "Re: Farm and Ranch Vacations Book." E-mail to the author. 17 October 2011.

Hull, Tim. *Arizona*. 10th ed. Berkeley, CA: Avalon Travel Publishing, 2008

Hurt, R. Douglas. *Indian Agriculture in America*. Lawrence: University Press of Kansas, 1987.

Jolly, Desmond A., and Kristin A. Reynolds. *Consumer Demand for Agricultural and On-Farm Nature Tourism*. Davis: University of California Small Farm Center, 2005.

Jones, Scottie. "Re: Chapter on Farm Stays." E-mail to the author. 15 January 2012.

———. "Re: Farm/Ranch Vacations Book." E-mail to the author. 10 October 2011.

Karanovitch, Fran. Personal interview. 12 February 2012.

Keating, Michelle. "Grace Miller Dude Duenna." *Tucson Daily Citizen*, 20 February 1971: 4.

Kilgore, Eugene. *Gene Kilgore's Ranch Vacations: The Complete Guide to Guest and Resort, Fly-fishing, and Cross-country Skiing ranches in the United States and Canada.* Santa Fe: John Muir Publications, 1999. Print.

Krueger, Christy. "Running Tanque Verde Ranch Resort is a Labor of Love." *Inside Tucson Business.* 28 December 2009: 10.

Lanning, Kimber. Personal interview. 11 September 2011.

———. "Re: Agritourism Chapter." E-mail to the author. 25 January 2012.

Leones, Julie, Douglas Dunn, Marshall Worden and Robert E. Call. "Agricultural Tourism in Cochise County, Arizona. Characteristics and Economic Impacts." Arizona Cooperative Extension. Tucson: University of Arizona, June 1994.

Littlefield, Joanne. "Direct Farm Marketing and Agri-tourism in Arizona." *Arizona Land and People* 49, no. 1 (2004): 10–11.

Loew, Laurel. Personal interview. 7 March 2011.

Loew, Stewart. "Re: Farm Vacations Book." E-mail to the author. 8 January 2012.

Malkin, Steve. Personal interview. 27 March 2012.

Mattern, Hal. "Circle Z Melds Ranch, Resort." *Arizona Republic*, 23 August 1992.

McKenzie, Michael. Personal interview. 7 November 2011.

———. "Re: Farm Vacations Book." E-mail to the author. 13 March 2011.

Mendelsohn, Deborah. "Re: Ranch and Farm Vacations Book." E-mail to the author. 8 February 2012.

Milburn, Betty. "Kenyon Guest Ranch." *Tucson Daily Citizen*. 3 May 1958: 40.

Milton, Corinne Holm. *Corona: Torero y Artista=Bullfighter & Artist*. Santa Fe, NM: Sunstone Press, 1989.

Murphree, Julie. "Re: Comments for farm stay chapter." E-mail to the author. 25 March 2012.

———. "Re: Farm and Ranch Vacations." E-mail to the author. 13 February 2012.

Nabhan, Gary. "Our Farm." 17 November. 2010. Accessed 14 January 2012. http://garynabhan.com/i/archives/806.

———. Personal interview. 19 May 2012.

Nabhan, Gary, and Kelley Watters. "Mom-and-pop vs. big-box stores in the food desert." *Grist Mag*, 1 June 2011.

Nett, Walt. "Triangle T Dodges the Bullet." *Arizona Daily Star.* 24 February 1991.

O'Neal, Bill. *Historic Ranches of the Old West.* Austin, TX: Eakin Press, 1997.

Otero Litel, Olga. "History of the Otero Family." *Mexican Holiday Press.* 5 May 1975.

Petroff, Janet. "Gallatin Canyon Will Remain Full of Grace." *Bozeman Daily Chronicle*, 2 November 1983: 29.

Potenza, Sarah. "Re: Book on Farm and Ranch Vacations." E-mail to the author. 12 September 2011.

Raneri, Stacy. Personal interview. 29 November 2011.

———. "Re: Farm and Ranch Vacations Book." E-mail to the author. 12 January 2012.

Record, Ian W. *Big Sycamore Stands Alone: the Western Apaches, Aravaipa, and the Struggle for Place.* Norman: University of Oklahoma Press, 2008.

Ross, Chris. "Riding for the Circle Z." *U-T San Diego*, 29 January 2012: G2.

Rugg, Lynne Thompson. Personal interview. 21 April 2012.

———. "Re: Kenyon Ranch." E-mail to the author. 24 April 2012.

Sayre, Nathan Freeman. *Ranching, Endangered Species and Urbanization in the Southwest: Species of Capital.* Tucson: University of Arizona Press, 2002.

Schultz, Veronica. "Re: Farm and Ranch Vacations Book." E-mail to the author. 10 December 2011.

Schwennesen, Eric. Personal interview. 5 December 2011.

———. "Re: Ranch Vacations Book." E-mail to the author. 9 January 2012.

Smith, Jeff. "The Crown C: Heaven on a Cattle Ranch." *Tucson Citizen*. 22 July 1982.

Soper, Jock. Personal interview. 12 February 2011.

———. "Re: Circle Z Ranch." E-mail to the author. 12 December 2011.

Steele, Carol. Personal interview. 24 March 2012.

———. "Re: Farm Vacations Book." E-mail to the author. 16 November 2011.

Sternberg, Chuck. Personal interview. 13 October 2011 and 18 November 2011.

Stewart, Janet Ann. *Arizona Ranch Houses: Southern Territorial Styles, 1867–1900.* Tucson: Arizona Historical Society, 1974.

Stover, Wendy. Personal interview. 12 February 2012.

———. "Re: Farm and Ranch Vacations Book." E-mail to the author. 20 March 2011.

Strittmatter, Janet H. "Rancho Linda Vista Historic District." National Register of Historic Places. United States Department of the Interior. 20 June 1998. Print.

Trimble, Marshall. *Arizona: A Cavalcade of History.* Tucson, AZ: Treasure Chest Publications. 1989.

———. *Arizona: A Panoramic History of a Frontier State.* Garden City, NY: Doubleday, 1977.

———. Personal interview. 2 November 2011.

Tronstad, Russell. Personal interview. 9 September 2011.

———. "Re: Farming History." E-mail to the author. 31 October 2011.

———. "Is a Bed and Breakfast Ranching?" *Arizona Ranchers' Management Guide*. R. Gum, G. Ruyle, and D. Rice (eds.). Tucson: Arizona Cooperative Extension. 1996.

Tronstad, Russell, Julie Leones. *Direct Farm Marketing and Tourism Handbook.* R. Tronstad and J. Leones (eds.). Tucson: Arizona Cooperative Extension. 2005.

True, Russell. Personal interview. 21 December 2011.

———. "Re: Farm and Ranch Vacations Book." E-mail to the author. 1 April 2011.

———. "Re: Farm and Ranch Vacations Book." E-mail to the author. 27 December 2011.

Volante, Ric. "Pinal Rejects Building 2,300 Homes Near Oracle." *Arizona Daily Star*, 27 March 1984.

———. "Ranch-Rezoning Request Omits Big-Game Preserve." *Arizona Daily Star*, 25 September 1983.

Weintraub, Vera. Personal interview. 28 December 2011.

———. "Re: Kenyon Ranch." E-mail to the author. 12 January 2012.

Williams, Dick. "Home on the Range at the Kenyon." *Southern Arizona Trails*, 13 September 1988.

Wooddancer, Janet. "Rancho De La Osa Draws You Back, Back…." *The Territorial* (Tucson, AZ) 9 April 1987: 15.

Woodruff, Susan. Personal interview. 15 January 2012.

———. "Re: Farm and Ranch Vacations." E-mail to the author. 4 January 2012.

Zoldan, Karyn. "Producing Community." *Tucson Weekly.* 3 July 2008: 1.

About the Author

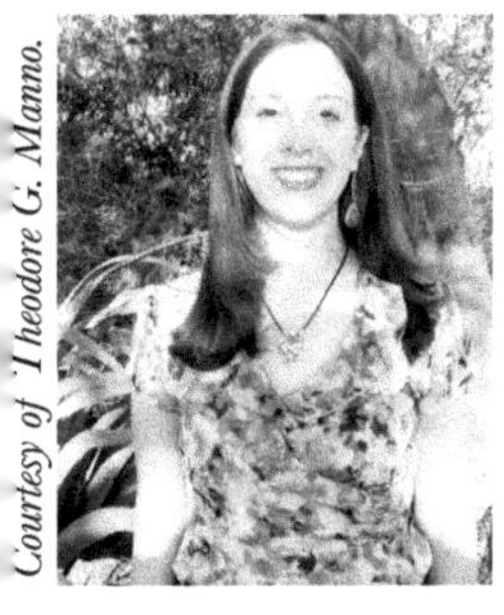
Courtesy of Theodore G. Manno.

A librarian and travel writer specializing in wildlife, the environment and historic preservation, Lili DeBarbieri's writing and photography have appeared in dozens of publications worldwide, including the *Los Angeles Times*, *Fine Books and Collections Magazine* and *Earth Island Journal*. Lili has traveled extensively in more than thirty countries, across five continents and over forty U.S. states. Based in Tucson, she is a stringer for Agence France-Presse and chair-elect of the Arizona Library Association's International Interest Group.

Visit us at
www.historypress.net

www.ingramcontent.com/pod-product-compliance
Lightning Source LLC
LaVergne TN
LVHW052341100826
845147LV00021B/1137
* 9 7 8 1 6 0 9 4 9 4 6 0 5 *